Stability Analysis of Neural Networks and Evolving Intelligent Systems

Jose de Jesus Rubio

Stability Analysis of Neural Networks and Evolving Intelligent Systems

Springer

Jose de Jesus Rubio
ESIME Azcapotzalco
Instituto Politecnico Nacional, Sección de
Estudios de Posgrado e Investigación
Ciudad de Mexico, Distrito Federal, Mexico

ISBN 978-3-031-87281-5 ISBN 978-3-031-87282-2 (eBook)
https://doi.org/10.1007/978-3-031-87282-2

This Springer imprint is published by the registered company Springer Nature Switzerland AG
The registered company address is: Gewerbestrasse 11, 6330 Cham, Switzerland

Preface

The goal of this book is to provide a detailed presentation of the stability of the neural networks and evolving intelligent systems. The neural networks and evolving intelligent systems are very interesting investigation fields, since they have many applications in the prediction and modeling of blending process, population process, brain signals, or eye signals.

The neural networks and the evolving intelligent systems are different in one characteristic; however, they are equal in other characteristic. A neural network has the ability to reorganize the model and adapt itself to a changing environment where the structure is static and the parameters learning is dynamic, while an evolving intelligent system has the ability to reorganize the model and adapt itself to a changing environment where both the structure and parameters learning are dynamic and are performed simultaneously. Therefore, the neural networks and the evolving intelligent systems are equal in that in both the parameters learning is dynamic, and they are different in that in one the structure is static and in the other the structure is dynamic.

This book is expected to be used primarily by researchers and secondarily by students and in the area of intelligent, control, robotic, energy, biological, mechanical, mechatronic, and computing systems.

The suggested use of the book is to be focused on each kind of intelligent system because they are different from each other, Chaps. 1–5 for the stability of the neural networks, Chaps. 6–10 for the stability of the evolving intelligent systems.

Mexico City, Mexico
2024

José de Jesús Rubio

Acknowledgments

The author thanks the Institute of Electrical and Electronics Engineers to let the re-use of portions of the work of [1,6] in Chaps. 1 and 6 of this work.

The author thanks the Elsevier B.V. to let the re-use of portions of the work of [4,5,7,8,10] in Chaps. 4, 5, 7, 8, and 10 of this work.

The author thanks the Springer-Verlag Berlin Heidelberg to let the re-use Of portions of the work of [2,3,9] in Chaps. 2, 3, and 9 of this work.

The author thanks the Instituto Politécnico Nacional, Consejo Nacional de Humanidades Ciencias y Tecnologías, Secretaría de Investigación y Posgrado, and Comisión de Operación y Fomento de Actividades Académicas for their help in this study.

References

1. J.J. Rubio, P. Angelov, J. Pacheco, Uniformly stable backpropagation algorithm to train a feedforward neural network. IEEE Trans. Neural Netw. **22**(3), 356–366 (2011)
2. J.J. Rubio, Analytic neural network model of a wind turbine. Soft Comput. **19**(12), 3455–3463 (2015)
3. J.J. Rubio, Interpolation neural network model of a manufactured wind turbine. Neural Comput. Appl. **28**(8), 2017–2028 (2017)
4. J.J. Rubio, I. Elias, D.R. Cruz, J. Pacheco, Uniform stable radial basis function neural network for the prediction in two mechatronic processes. Neurocomputing **227**, 122–130 (2017)
5. J.J. Rubio, USNFIS: uniform stable neuro fuzzy inference system. Neurocomputing **262**, 57–66 (2017)
6. J.J. Rubio, SOFMLS: online self-organizing fuzzy modified least square network. IEEE Trans. Fuzzy Syst. **17**(6), 1296–1309 (2009)

7. J.J. Rubio, J.H. Perez Cruz, Evolving intelligent system for the modelling of nonlinear systems with dead-zone input. Appl. Soft Comput. **14**(Part B), 289–304 (2014)
8. J.J. Rubio, Evolving intelligent algorithms for the modelling of brain and eye signals. Appl. Soft Comput. **14**(Part B), 259–268 (2014)
9. J.J. Rubio, A. Bouchachia, MSAFIS: an evolving fuzzy inference system. Soft Comput. **21**(9), 2357–2366 (2017)
10. J.J. Rubio, Error convergence analysis of the SUFIN and CSUFIN. Appl. Soft Comput. **72**, 587–595 (2018)

Contents

Chapter 1
Stability Analysis of Neural Networks and Evolving Intelligent Systems

1 Introduction

A neural network has the ability to reorganize the model and adapt itself to a changing environment where the structure is static and the parameters learning is dynamic, while an evolving intelligent system has the ability to reorganize the model and adapt itself to a changing environment where both the structure and parameters learning are dynamic and are performed simultaneously. The stable neural networks and stable evolving intelligent systems are the models where their structure, weights, and parameters remain bounded through the time. The neural networks and evolving intelligent systems are applied to many online fields, but the stability of the neural networks and evolving intelligent systems is not always assured, and it could damage the devices causing accidents. Therefore, it would be interesting to assure the stability of the neural networks and evolving intelligent systems.

The stable algorithms utilized in the neural networks and evolving intelligent systems must satisfy three conditions to assure their stability in the learning: They need to be compact, they need to be effective, and they need to be stable. The neural networks and evolving intelligent systems have only one hidden layer to assure their compactness. The neural networks and evolving intelligent systems are in discrete time, where one analysis based on the Lyapunov method is considered to assure the stability for the modeling error; additionally, other analysis as a consequence of the Lyapunov method is considered to assure the boundedness of the weights and parameters.

This book contains two parts: Part 1 of Chaps. 1–5 contains the stability analysis of neural networks, and part 2 of Chaps. 6–10 contains the stability analysis of evolving intelligent systems. In this book, the stability analysis of the neural networks and evolving intelligent systems is mainly obtained by the Lyapunov method. In this book, the neural networks are applied in the prediction of the distribution of loads in a warehouse, in the modeling of the wind turbine behavior,

J. de J. Rubio, *Stability Analysis of Neural Networks and Evolving Intelligent Systems*, https://doi.org/10.1007/978-3-031-87282-2_1

in the modeling of the crude oil blending process, and in the modeling of the beetle population process, and the evolving intelligent systems are applied in modeling of brain signals, in the modeling of eye signals, in the modeling of nonlinear systems with dead-zone input, and in the modeling of the Box Jenkins furnace.

The detailed description of chapters in this book is as follows:

In Chap. 1, a backpropagation algorithm is introduced for the learning of a neural network. The major contributions of this chapter are as follows: (1) A theorem to assure the uniform stability of the general discrete-time systems is proposed, (2) it is proven that the backpropagation algorithm with a new time varying rate is uniformly stable for online identification, and the identification error converges to a small zone bounded by an uncertainty, (3) it is proven that the weights' error is bounded by the initial weights' error, i.e., the overfitting is not presented in the proposed algorithm, (4) the backpropagation is applied to predict the distribution of loads that a transelevator receives from a trailer and places in the deposits each hour in a warehouse, and the deposits in the warehouse can be reserved in advance using the prediction results, (5) the backpropagation algorithm is compared with the recursive least square algorithm and the Sugeno fuzzy inference system in the problem of the prediction of the distribution of loads in a warehouse, giving that the first and the second are stable and the third is unstable, and (6) the backpropagation algorithm is compared with the recursive least square algorithm and the Kalman filter algorithm in an academic example. This work is also published in [1].

In Chap. 2, an analytic neural network model is introduced for the modeling of the wind turbine behavior. The proposed hybrid method is the combination of the analytic and neural network models. The neural network model is used as a compensator to improve the approximation of the analytic model. It is guaranteed that the error of the analytic neural network model is smaller than the error of the analytic model. Two experiments show the effectiveness of the proposed technique. This work is also published in [2].

In Chap. 3, an interpolation neural network is introduced for the learning of a wind turbine behavior with incomplete data. The proposed hybrid method is the combination of an interpolation algorithm and a neural network. The interpolation algorithm is applied to estimate the missing data of all the variables; later, the neural network is employed to learn the output behavior. The proposed method avoids the requirement to know all the system data. Experiments show the effectiveness of the proposed technique. This work is also published in [3].

In Chap. 4, a method to obtain a stable algorithm is presented for the learning of a radial basis function neural network. The method consists of the following processes: (1) the radial basis function neural network is linearized, (2) the algorithm for the learning of the radial basis function neural network is introduced, (3) stability of the mentioned technique is assured, (4) convergence of the suggested method is guaranteed, and (5) boundedness of parameters in the focused technique is assured. The abovementioned method is applied for the learning of two mechatronic processes. This work is also published in [4].

In Chap. 5, a stable neuro fuzzy inference system is designed from the multilayer neural network and fuzzy inference system to satisfy the three conditions for the big data learning: (1) It utilizes the numerator of the average defuzzifier instead of the average defuzzifier to be compact, (2) it employs Gaussian functions instead of sigmoid functions to be effective, and (3) it uses a time varying learning speed instead of the constant learning speed to be stable. The suggested technique is applied for the modeling of the crude oil blending process and the beetle population process. This work is also published in [5].

In Chap. 6, an online self-organizing fuzzy modified least square (SOFMLS) network is proposed. The network generates a new rule, if the smallest distance between the new data and all the existing rules (the winner rule) is more than a prespecified radius. The major contributions of this chapter are as follows: (1) A new network is proposed. In this network, unidimensional membership functions are used, and only two parameters for each rule are employed, thus reducing the number of parameters. The network avoids the singularity produced by the widths in the antecedent part for online learning. (2) A new pruning algorithm based on the density is proposed, where the density is the number of times that each rule is used in the algorithm. The rule that has the smallest density (the looser rule) in a selected number of iterations is pruned if the value of its density is smaller than a prespecified threshold. (3) The stability of the proposed algorithm is proven, and the bound for the average of the identification error is found. The condition that led the algorithm to avoid the local minimum is found, and it is proven that the parameters error is bounded by the initial parameters error. Three simulations give the effectiveness of the suggested algorithm. This work is also published in [6].

In Chap. 7, the modeling problem of nonlinear systems with dead-zone input is considered. To solve this problem, an evolving intelligent system is proposed. The uniform stability of the modeling error for the aforementioned system is guaranteed by means of a Lyapunov-like analysis. The effectiveness of the proposed technique is verified by simulations. This work is also published in [7].

In Chap. 8, the modeling problem of brain and eye signals is considered. To solve this problem, three important evolving and stable intelligent algorithms are applied: the sequential adaptive fuzzy inference system (SAFIS), uniform stable backpropagation algorithm (SBP), and online SOFMLS networks. The effectiveness of the studied methods is verified by simulations. This work is also published in [8].

In Chap. 9, the problem of learning in big data is considered. To solve this problem, a new algorithm is proposed as the combination of two important evolving and stable intelligent algorithms: the SAFIS and stable gradient descent algorithm (SGD). The modified sequential adaptive fuzzy inference system (MSAFIS) is the SAFIS with the difference that the SGD is used instead of the Kalman filter for updating parameters. The SGD improves the Kalman filter because the first obtains a better learning in big data. The effectiveness of the introduced method is verified by two experiments. This work is also published in [9].

In Chap. 10, the error convergence of the SAFIS and the MSAFIS is analyzed. SAFIS utilizes the extended Kalman filter, while MSAFIS uses the gradient descent technique. First, proposed algorithms are linearized to get their modeling dynamic equations. Second, Lyapunov strategy is utilized to ensure the error convergence of studied networks. Two examples show the performance of advised algorithms. This work is also published in [10].

References

1. J.J. Rubio, P. Angelov, J. Pacheco, Uniformly stable backpropagation algorithm to train a feedforward neural network. IEEE Trans. Neural Netw. **22**(3), 356–366 (2011)
2. J.J. Rubio, Analytic neural network model of a wind turbine. Soft. Comput. **19**(12), 3455–3463 (2015)
3. J.J. Rubio, Interpolation neural network model of a manufactured wind turbine. Neural Comput. Appl. **28**(8), 2017–2028 (2017)
4. J.J. Rubio, I. Elias, D.R. Cruz, J. Pacheco, Uniform stable radial basis function neural network for the prediction in two mechatronic processes. Neurocomputing **227**, 122–130 (2017)
5. J.J. Rubio, USNFIS: uniform stable neuro fuzzy inference system. Neurocomputing **262**, 57–66 (2017)
6. J.J. Rubio, SOFMLS: online self-organizing fuzzy modified least square network. IEEE Trans. Fuzzy Syst. **17**(6), 1296–1309 (2009)
7. J.J. Rubio, J.H. Perez Cruz, Evolving intelligent system for the modelling of nonlinear systems with dead-zone input. Appl. Soft Comput. **14**(Part B), 289–304 (2014)
8. J.J. Rubio, Evolving intelligent algorithms for the modelling of brain and eye signals. Appl. Soft Comput. **14**(Part B), 259–268 (2014)
9. J.J. Rubio, A. Bouchachia, MSAFIS: an evolving fuzzy inference system. Soft Comput. **21**(9), 2357–2366 (2017)
10. J.J. Rubio, Error convergence analysis of the SUFIN and CSUFIN. Appl. Soft Comput. **72**, 587–595 (2018)

Chapter 2
A Uniformly Stable Backpropagation Algorithm to Train a Feedforward Neural Network

1 Introduction

The online neural networks can be used in many fields, including nonlinear adaptive control, fault detection, diagnostics, performance analysis of dynamic systems, pattern and image recognition, time-series, identification of nonlinear systems, intelligent agents, modeling, robotic, and mechatronic systems. The stability problem of neural networks is important for the aforementioned online fields, and the stability of the neural networks is not always assured.

There are some researchers who have worked with the stability of continuous time neural networks as are [1–9].

In [1], they study the approximation and the learning properties of one class of recurrent networks, known as high-order neural networks, and they apply these architectures to the identification of dynamic systems. In [2], the stability conditions of online identification are derived by Lyapunov-Krasovskii approach, which are described by linear matrix inequality. In [3], they present the sufficient conditions for the global asymptotic stability for a kind of recurrent neural network. In [4], they consider the robust stability of neural networks with multiple delays. The work of [5] is concerned with the global robust exponential stability of a class of interval Cohen-Grossberg neural networks with both multiple time varying delays and continuously distributed delays. In [6], the static neural network model and a local field neural network model are theoretically compared in terms of their trajectory transformation property, equilibrium correspondence property, nontrivial attractive manifold property, global convergence, as well as stability in many different senses. In [7], dynamic multilayer neural networks are used for nonlinear system online identification, and the passivity approach is applied to access several stability properties of the neuro-identifier. In [8], the passivity-based approach is used to derive stability conditions for dynamic neural networks with different time scales. In [9], the Lyapunov function approach is used to rigorously analyze the convergence of weights, with the use of the backpropagation algorithm, toward minima of the

J. de J. Rubio, *Stability Analysis of Neural Networks and Evolving Intelligent Systems*, https://doi.org/10.1007/978-3-031-87282-2_2

error function. All the works are interesting, but all consider the continuous-time neural networks, and there are some systems that are better described in discrete time, for example, the population systems of some kind of animals [10], or the annual expenses in an industry [11], or the interest earned by the loan of a bank [11], or the prediction of the distribution of loads received each hour in a warehouse, that is way it is important to consider the stability of the discrete-time neural networks.

There are some researchers who have worked with the stability of discrete-time neural networks as are [12–17].

In [12], a double dead-zone is used to assure the stability of the identification error in the gradient descent algorithm. In [13], they derive a condition for robust local stability of the multilayer recurrent neural networks. In [14], an input to state stability approach is used to create robust training algorithms for discrete-time neural networks. The work of [15] suggests new learning laws for Mamdani and Takagi-Sugeno-Kang type fuzzy neural networks based on input-to-state stability approach. In [16], the input-to-state stability approach is applied to access robust training algorithms of discrete-time recurrent neural networks. In [17], they modify the backpropagation approach, and they employ a time varying rate that is determined from the input-output data and the model structure and stable learning algorithms for the premise and the consequence parts of the fuzzy rules are proposed. All the works propose new neural network algorithms as in [13], or they modify the general backpropagation employing a time varying rate to prove the input-to-state stability as in [12, 14–17]; in this chapter it is proven that the backpropagation algorithm with a new time varying rate is uniformly stable.

On the other hand, there is some research related with the warehouses as is [18–22].

The authors in [18] propose a method for selecting and materializing views, which selects and horizontally fragments a view and recomputes the size of the stored partitioned view while deciding further views to select. In [19], they consider a matrix-based discrete event control approach for a warehouse, and the control system is organized in two modules: a dynamic model and a controller. In [20], they focus on the technical challenges of designing and implementing an effective data warehouse for health care information. In [21], they propose, as an extension to the data warehouse model, knowledge warehouse architecture that will not only facilitate the capturing and coding of knowledge but also enhance the retrieval and sharing of knowledge across the organization. In [22], they propose a new constrained evolutionary algorithm for the maintenance-cost view-selection problem. All the works are interesting, but none uses the neural networks for the prediction of the distribution of loads in a warehouse, and in [21], they only mention that it could be made.

In this chapter, it is proposed a theorem to assure the uniform stability of the discrete-time systems, it is proven that the backpropagation algorithm with a new time varying rate is uniformly stable for online identification, the identification error converges to a small zone bounded by the uncertainty, and the weights' error is bounded by the initial weights' error; the backpropagation is applied to predict the distribution of loads that a transelevator receives from a trailer and places in the deposits each hour in a warehouse, the deposits in the warehouse can be reserved in advance using the prediction results, the backpropagation algorithm is compared with the recursive least square algorithm and with the Sugeno fuzzy inference system in the problem of the prediction of the distribution of loads inside a warehouse, and the backpropagation algorithm is compared with the recursive least square and with the Kalman filter in an academic example.

This chapter is organized as follows. In Sect. 2, the theorem that proves the uniformly stability of the discrete-time systems is presented. In Sect. 3, the general backpropagation to train a feedforward neural network with a hidden layer is presented. In Sect. 4, the uniform stability of the backpropagation algorithm is proven. In Sect. 5, the application of the proposed algorithm is described. In Sect. 6, a brief description of the warehouse is presented. In Sect. 7, the backpropagation algorithm is compared with the recursive least square algorithm, with the Sugeno fuzzy inference system, and with the Kalman filter algorithm in the problem of the prediction of the distribution of loads in a warehouse and in an academic example. Finally, in Sect. 8, the results and the possible future research are explained.

2 Preliminaries

Let us consider the following discrete-time nonlinear system:

$$x_{k+1} = f\left[x_k, u_k\right], \tag{2.1}$$

where $u_k \in \Re^m$ is the input vector, $x_k \in \Re^n$ is the state vector, and u_k and x_k are known. f is an unknown nonlinear smooth function $f \in C^\infty$.

Definition 2.1 The system (2.1) is said to be uniformly stable if $\forall \epsilon > 0$, $\exists\, \delta = \delta(\epsilon)$ such that

$$\|x_{k1}\| < \delta \Rightarrow \|x_k\| < \epsilon, \quad \forall k > k_1. \tag{2.2}$$

If the system has $\delta = \delta(\epsilon, k)$, then the system (2.1) only is stable.

Now, a basic stability theorem for discrete-time nonlinear systems is given, it is an analogous version of the continuous-time version given by Byrnes et al. [23] and of the delayed continuous-time version given by Rubio and Yu [12].

Theorem 2.1 *Let $L_k(x(k))$ be a Lyapunov function of the discrete-time nonlinear system (2.1), if it satisfies*

$$\begin{aligned} \gamma_1\left(\|x_k\|\right) \le L_k(x_k) \le \gamma_2\left(\|x_k\|\right), \\ \Delta L_k(x_k) \le -\gamma_3\left(\|x_k\|\right) + \gamma_3\left(\delta\right), \end{aligned} \tag{2.3}$$

where δ is a positive constant, $\gamma_1(\cdot)$ and $\gamma_2(\cdot)$ are K_∞ functions, and $\gamma_3(\cdot)$ is a K function; then the system (2.1) is uniformly stable.

Proof *See [24] for the proof.*

3 The Backpropagation Algorithm to Train a Neural Network

Let us consider the following unknown discrete-time nonlinear system:

$$y(k) = f\left[X_k\right], \tag{2.4}$$

where $X_k = \left[x_1(k)\ldots, x_i(k), \ldots, x_N(k)\right]^T = \left[y(k-1), \ldots, y(k-n), u\left(k-1\right), \ldots, u\left(k-m\right)\right]^T \in \Re^{N\times 1} (N = n+m)$ is the input vector, $u(k-1) \in \Re$ is the input of the plant, $y(k) \in \Re$ is the output of the plant, and f is an unknown nonlinear function, $f \in C^\infty$. The output of the neural network with one hidden layer can be expressed as [25–27]

$$\begin{aligned} \widehat{y}(k) &= V_k \Phi_k = \sum_{j=1}^{M} V_{jk}\phi_{jk}, \\ \Phi_k &= \left[\phi_{1k}, \ldots, \phi_{jk}, \ldots, \phi_{Mk}\right]^T, \\ \phi_{jk} &= \tanh\left(\sum_{i=1}^{N} W_{ijk}x_i(k)\right), \end{aligned} \tag{2.5}$$

where $i = 1, \ldots, N$, $j = 1, \ldots, M$, $X_k \in \Re^{N\times 1}$ is the input vector given by (2.4), $\widehat{y}(k) \in \Re$ is the output of the neural network, $V_k \in \Re^{1\times M}$ and $W_k \in \Re^{M\times N}$ are the weights of the output and the hidden layer of the neural network, respectively, $W_{ijk} \in \Re$, $x_i(k) \in \Re$, $\Phi_k \in \Re^{M\times 1}$, $\phi_{jk} \in \Re$, $V_{jk} \in \Re$, and Fig. 2.1 shows the feedforward neural network.

4 Stability of the Backpropagation Algorithm

The stability of the parameter learning is needed, because this algorithm works online. First, the model is linearized, and later, the stability of the proposed algorithm is analyzed.

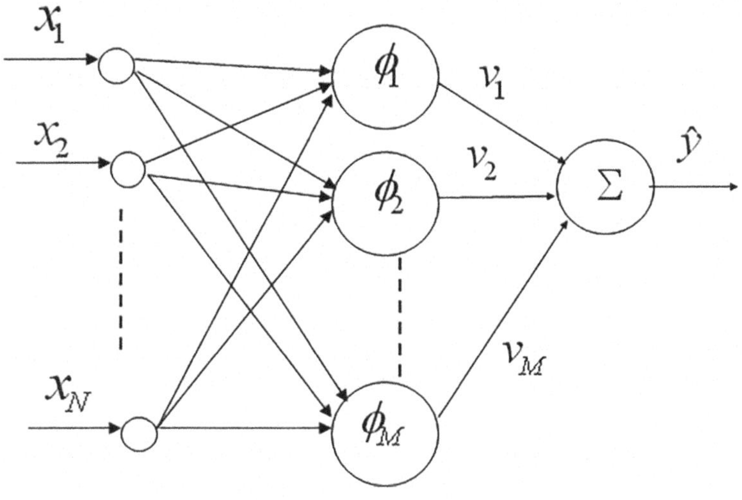

Fig. 2.1 Architecture of the neural network

According to the Stone-Weierstrass theorem [28], the unknown nonlinear function f of (2.4) is approximated as

$$\begin{aligned} y(k) &= V_* \Phi_{*k} + \in_f = \sum_{j=1}^{M} V_{j*} \phi_{*jk} + \in_f, \\ \Phi_{*k} &= \left[\phi_{*1k}, \ldots, \phi_{*jk}, \ldots, \phi_{*Mk}\right]^T, \\ \phi_{*jk} &= \tanh\left(\sum_{i=1}^{N} W_{ij*} x_i(k)\right), \end{aligned} \tag{2.6}$$

where $\Phi_{*k} \in \Re^{M\times 1}$, $\in_f = y(k) - V_* \Phi_{*k} \in \Re$ is the modeling error, $\phi_{*jk} \in \Re$, $V_{j*} \in \Re$, $W_{ij*} \in \Re$, $V_{j*} \in \Re$, and $W_{ij*} \in \Re$ are the optimal parameters that can minimize the modeling error $\in_f$ [1].

First, the network model is linearized, and it is used to define the parameters updating and to prove the stability of the proposed algorithm.

In the case of two independent variables, a function has a Taylor series as follows:

$$\begin{aligned} f(\omega_1, \omega_2) &= f(\omega_{1^0}, \omega_{2^0}) + \left(\omega_1 - \omega_{1^0}\right) \tfrac{\partial f(\omega_1,\omega_2)}{\partial \omega_1} \\ &+ \left(\omega_2 - \omega_{2^0}\right) \tfrac{\partial f(\omega_1,\omega_2)}{\partial \omega_2} + R_f, \end{aligned} \tag{2.7}$$

where $R_f \in \Re$ is the remainder of the Taylor series. If we let ω_1 and ω_2 correspond to $W_{ijk} \in \Re$ and $V_{jk} \subset \Re$ and ω_{1^0} and ω_{2^0} correspond to $W_{ij*} \in \Re$ and $V_{j*} \in \Re$ and let us define $\widetilde{W}_{ijk} = W_{ijk} - W_{ij*} \in \Re$ and $\widetilde{V}_{jk} = V_{jk} - V_{j*} \in \Re$, then the

Taylor series is applied to linearize (2.5) as follows [2, 12, 29]:

$$V_k\Phi_k = V_*\Phi_{*k} + \sum_{j=1}^{M}\sum_{i=1}^{N}\widetilde{W}_{ijk}\frac{\partial V_k\Phi_k}{\partial W_{ijk}} + \sum_{j=1}^{M}\widetilde{V}_{jk}\frac{\partial V_k\Phi_k}{\partial V_{jk}} + R_f, \tag{2.8}$$

where $\frac{\partial V_k\Phi_k}{\partial W_{ijk}} \in \Re$ and $\frac{\partial V_k\Phi_k}{\partial V_{jk}} \in \Re$; please note that $V_k\Phi_k = \sum_{j=1}^{M} V_{jk}\phi_{jk}$ and $V_*\Phi_{*k} = \sum_{j=1}^{M} V_{j*}\phi_{*jk}$. As all the parameters are scalars, the Taylor series is well applied. Considering (2.5) and using the chain rule [2, 12, 29–31] give

$$\begin{aligned}\frac{\partial V_k\Phi_k}{\partial W_{jk}} &= V_{jk}\frac{\partial\Phi_k}{\partial W_{jk}} = V_{jk}\frac{\partial \tanh\left(\sum_{i=1}^{N} W_{ijk}x_i(k)\right)}{\partial W_{ijk}} \\ &= V_{jk}\mathrm{sech}^2\left(\sum_{i=1}^{N} W_{ijk}x_i(k)\right)x_i(k) = \sigma_{ijk},\end{aligned} \tag{2.9}$$

where $\sigma_{ijk} = V_{jk}\mathrm{sech}^2(\sum_{i=1}^{N} W_{ijk}x_i(k))x_i(k) \in \Re$ because $V_{jk} \in \Re$, $\mathrm{sech}^2(\sum_{i=1}^{N} W_{ijk}x_i(k)) \in \Re$ and $x_i(k) \in \Re$.

$$\frac{\partial V_k\Phi_k}{\partial V_{jk}} = \frac{\partial \sum_{j=1}^{M} V_{jk}\phi_{jk}}{\partial V_{jk}} = \phi_{jk}, \tag{2.10}$$

where $\phi_{jk} = \tanh(\sum_{i=1}^{N} W_{ijk}x_i(k)) \in \Re$. Substituting $\frac{\partial V_k\Phi_k}{\partial W_{ijk}}$ of (2.9) and $\frac{\partial V_k\Phi_k}{\partial V_{jk}}$ of (2.10) into (2.8) gives

$$\begin{aligned}V_k\Phi_k = V_*\Phi_{*k} &+ \sum_{j=1}^{M}\sum_{i=1}^{N}\widetilde{W}_{ijk}\sigma_{ijk} \\ &+ \sum_{j=1}^{M}\widetilde{V}_{jk}\phi_{jk} + R_f.\end{aligned} \tag{2.11}$$

Let us define the identification error $e(k) \in \Re$ as follows:

$$e(k) = \widehat{y}(k) - y(k), \tag{2.12}$$

where $y(k)$ and $\widehat{y}(k)$ are defined in (2.4) and (2.5), respectively. Substituting (2.5), (2.6), and (2.11) into (2.12) gives

$$e\,(k) = \sum_{j=1}^{M} \widetilde{V}_{jk}\phi_{jk} + \sum_{j=1}^{M}\sum_{i=1}^{N} \widetilde{W}_{ijk}\sigma_{ijk} + \mu(k), \tag{2.13}$$

where $\mu(k) = R_f - \in_f$.

From (2.13), it is obtained that

$$\sum_{j=1}^{M} \widetilde{V}_{jk}\phi_{jk} + \sum_{j=1}^{M}\sum_{i=1}^{N} \widetilde{W}_{ijk}\sigma_{ijk} = e\,(k) - \mu(k). \tag{2.14}$$

The proposed backpropagation algorithm uses a new time varying rate as follows:

$$\begin{aligned} V_{jk+1} &= V_{jk} - \alpha_k \phi_{jk} e(k), \\ W_{ijk+1} &= W_{ijk} - \alpha_k \sigma_{ijk} e(k), \end{aligned} \tag{2.15}$$

where the new time varying rate α_k is

$$\alpha_k = \frac{\alpha_0}{2\left(\frac{1}{2} + \sum_{j=1}^{M} \phi_{jk}^2 + \sum_{j=1}^{M}\sum_{i=1}^{N} \sigma_{ijk}^2\right)},$$

where $i = 1, \ldots, N$, $j = 1, \ldots, M$, $\sigma_{ijk} = V_{jk}\mathrm{sech}^2(\sum_{i=1}^{N} W_{ijk}x_i(k))x_i(k) \in \Re$ is defined in (2.9), $\phi_{jk} = \tanh(\sum_{i=1}^{N} W_{ijk}x_i(k)) \in \Re$ is defined in (2.5) and used in (2.10), $e(k)$ is defined in (2.12), $0 < \alpha_0 \leq 1 \in \Re$, so $0 < \alpha_k \in \Re$, and it is assumed that the uncertainty is bounded [1, 2, 12, 15, 29, 32–35], where $\overline{\mu}$ is the upper bound of the uncertainty $\mu(k)$, $|\mu(k)| < \overline{\mu}$.

Remark 2.1 Please note that $e(k) = \widehat{y}(k) - y(k) = \sum_{j=1}^{M} V_{jk}\phi_{jk} - y(k)$ used in (2.15) is well defined because V_{jk}, ϕ_{jk}, and $y(k)$ are known.

The following theorem gives the stability of the proposed backpropagation algorithm.

Theorem 2.2 *The backpropagation algorithm (2.5), (2.12), and (2.15) applied for the identification of the nonlinear system (2.4) is uniformly stable, and the upper bound of the average identification error $e_p^2(k)$ satisfies*

$$\limsup_{T\to\infty}\frac{1}{T}\sum_{k=2}^{T} e_p^2(k) \le \alpha_0 \overline{\mu}^2, \tag{2.16}$$

where $e_p^2(k) = \frac{\alpha_k}{2} e^2(k-1)$, $0 < \alpha_0 \le 1 \in \Re$, and $0 < \alpha_k \in \Re$ are defined in (2.15), $e(k)$ is defined in (2.12), $\overline{\mu}$ is the upper bound of the uncertainty $\mu(k)$, $|\mu(k)| < \overline{\mu}$.

Proof See [24] for the proof.

Remark 2.2 There are two conditions to apply this algorithm for nonlinear systems: the first one is that the nonlinear system may have the form described by Eq. (2.4), and the second one is that the uncertainty $\mu(k)$ may be bounded.

Remark 2.3 The value of the parameter $\overline{\mu}$ is unimportant, because this parameter is not used in the algorithm. The bound of $\mu(k)$ is needed to guarantee the stability of the algorithm, but it is not used in the backpropagation algorithm (2.5), (2.12), (2.15).

Remark 2.4 The fact that $\mu(k)$ is bounded has been used for other authors in some earlier studies as are [1, 33, 34], and [35] in continuous-time systems and [2, 12, 15, 29], and [32] in discrete-time systems.

The following theorem proves that the weights of the proposed backpropagation algorithm are bounded.

Theorem 2.3 *When the average error $e_p^2(k)$ is bigger than the uncertainty $\alpha_0\overline{\mu}^2$, the weights' error is bounded by the initial weights' error as follows:*

$$\begin{aligned} & e_p^2(k) \ge \alpha_0 \overline{\mu}^2 \\ \Longrightarrow & \sum_{j=1}^{M} \widetilde{V}_{jk+1}^2 + \sum_{j=1}^{M}\sum_{i=1}^{N} \widetilde{W}_{ijk+1}^2 \le \sum_{j=1}^{M} \widetilde{V}_{j1}^2 + \sum_{j=1}^{M}\sum_{i=1}^{N} \widetilde{W}_{ij1}^2, \end{aligned} \tag{2.17}$$

where $i = 1, \ldots, N$, $j = 1, \ldots, M$, $\widetilde{V}_{jk}$ and $\widetilde{W}_{ijk}$ are defined in (2.7), $\widetilde{V}_{j1}$ and $\widetilde{W}_{ij1}$ are the initial weights' error, $e_p^2(k) = \frac{\alpha_k}{2} e^2(k)$, V_{jk+1}, W_{ijk+1}, $0 < \alpha_0 \le 1 \in \Re$, and $0 < \alpha_k \in \Re$ are defined in (2.15), $e(k)$ is defined in (2.12), $\overline{\mu}$ is the upper bound of the uncertainty $\mu(k)$, $|\mu(k)| < \overline{\mu}$.

Proof *See [24] for the proof.*

Remark 2.5 From Theorem 2.2 the average identification error $e_p^2(k)$ of the backpropagation algorithm is bounded, and from Theorem 2.3 the weights' error $\widetilde{V}_{jk+1}^2$ and $\widetilde{W}_{ijk+1}^2$ is bounded, i.e., the proposed backpropagation algorithm to train a feedforward neural network is uniformly stable in the presence of unknown and bounded uncertainties, and the overfitting mentioned in [14] and [27] is not presented. In addition, the identification error converges to a small zone bounded by the uncertainty $\overline{\mu}$.

5 The Proposed Algorithm

The proposed algorithm is as follows:

(1) Obtain the output of the nonlinear system $y(k)$ with Eq. (2.4), note that the nonlinear system may have the structure with Eq. (2.4), and the parameter N is selected according to this nonlinear system.
(2) Select the following parameters: V_1 and W_1 are selected as random numbers between 0 and 1. M is selected as an entire number, and α_0 is selected with positive values smaller than or equal to 1; obtain the output of the neural network $\widehat{y}(1)$ with Eq. (2.5).
(3) For each iteration k, obtain the output of the neural network $\widehat{y}(k)$ with Eq. (2.5), also obtain the identification error $e(k)$ with Eq. (2.12), and update the parameters V_{jk+1} and W_{ijk+1} with Eq. (2.15).
(4) Note that the behavior of the algorithm could be improved by changing the values of M or α_0.

Remark 2.6 The proposed neural network has one hidden layer. Some earlier results [1, 28], and [31] mention that there is a result where the feedforward neural network with one hidden layer is enough to approximate any nonlinear system.

6 The Warehouse

An automatic warehouse has elements used to make easy the work of moving loads from one place to another one in an automatic way. The loads are some objects inside of boxes that are saved in the warehouse until they are sent to the costumers. The deposits are the place where the loads are placed. Figure 2.2 shows the automatic warehouse in gray color, the deposits in black color, and the loads in brown color.

A transelevator moves inside the warehouse. This transelevator can be used to move some load from one place to another one in the warehouse, for example, from the floor to the deposit, from the deposit to the floor, from one deposit to another one, or from a trailer to the deposits. Figure 2.3 shows a transelevator inside the warehouse in yellow color, and Fig. 2.4 shows the same transelevator moving a load.

Figure 2.5 shows the trailer with the loads that are saved in the warehouse. The transelevator takes the loads from the trailer and places them in the deposits.

In this chapter, the main prediction problem in the warehouse is the distribution of the loads that the transelevator receives from the trailer and places in the deposits

Fig. 2.2 The automatic warehouse

Fig. 2.3 The transelevator inside the warehouse

each hour inside the warehouse, and the deposits in the warehouse can be reserved in advance using the prediction results.

7 Simulations

In this section, two examples are considered. In the first example, the backpropagation algorithm is applied for the prediction of the distribution of loads inside a warehouse, and the proposed algorithm is compared with the recursive least square

Fig. 2.4 The transelevator with a load

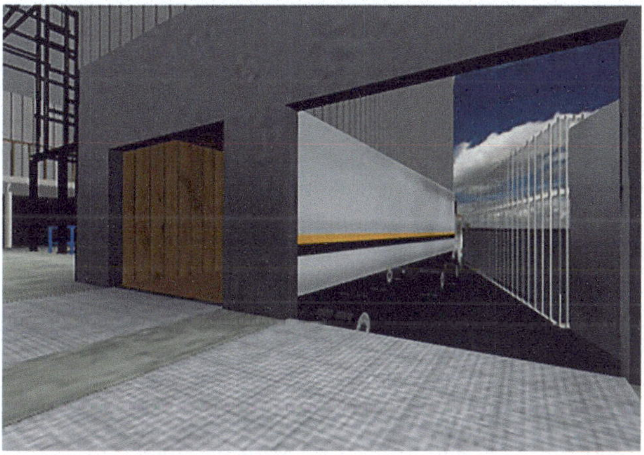

Fig. 2.5 The trailer with loads for the warehouse

algorithm given by Goodwin and Sin [36] and used by Angelov and Filev [37] and Kasabov and Song [38] and with the Sugeno fuzzy inference system given by Jang and Sun [27] and Wang [31]. In the second example, the backpropagation algorithm is applied in an academic problem, and the proposed algorithm is compared with the recursive least square algorithm given by Goodwin and Sin [36] and used by Angelov and Filev [37] and Kasabov and Song [38] and with the Kalman filter algorithm given by Haykin [25] and Goodwin and Sin [36] and used by Rubio and Yu [2].

The root mean square error (RMSE) [39] is used, and it is given as follows:

$$\mathrm{RMSE} = \left(\frac{1}{N} \sum_{k=1}^{N} e^2(k) \right)^{\frac{1}{2}}. \tag{2.18}$$

Example 2.1 In this example, the backpropagation is applied for the prediction of the distribution of loads that the transelevator receives from the trailer and places in the deposits each hour in the warehouse, there are three kinds of loads received by the transelevator inside the warehouse, these three kinds of loads are denoted as A, B, and C, and they are received in the warehouse each hour; the number of loads of kind A received each hour in the warehouse can vary from 4 to 5, the number of loads of kind B received each hour in the warehouse can vary from 3 to 4, and the number of loads of kind C received each hour in the warehouse can vary from 1 to 3. The data of 1800 hours are used for the training, and the data of the least 200 hours are used for the testing; the prediction is obtained with 200 hours in advance. Three neural networks are used for the training, and the same neural networks are used for the testing; $B(k)$ and $C(k)$ are the inputs and $A(k+200)$ is the output for the training of the first neural network, $A(k)$ and $C(k)$ are the inputs and $B(k+200)$ is the output for the training of the second neural network, and $A(k)$ and $B(k)$ are the inputs and $C(k+200)$ is the output for the training of the third neural network. Similar inputs are used for the testing of the three neural networks, and the outputs are not used for the testing. In this prediction example, the backpropagation algorithm is given as (2.5), (2.12), and (2.15) changing $y(k)$ by $y(k+200)$ and changing $e(k)$ by $e(k+200)$ [36]. The parameters of the backpropagation algorithm are $N = 2$, $M = 5$, V_{j1} and W_{ij1} are random number between 0 and 1, and $\alpha_0 = 1$. The backpropagation algorithm is compared with the recursive least square algorithm given by Goodwin and Sin [36] and used by Angelov and Filev [37] and Kasabov and Song [38] with parameters $P_1 = cI \in \Re^{2\times 2}$, where $c = 1$, V_1 is a random number between 0 and 1 and is compared with the Sugeno fuzzy inference system given by Jang and Sun [27] and Wang [31] with parameters $M = 2$, m_1, σ_1, and v_1 are random numbers between 0 and 1. The training results are shown in Fig. 2.6, and the testing results are shown in Fig. 2.7. Table 2.1 shows the training and the testing RMSE results using (2.18). Figure 2.8 shows that in this example not all the algorithms are stable because the Sugeno fuzzy inference system is not stable, and it is reported in Table 2.1.

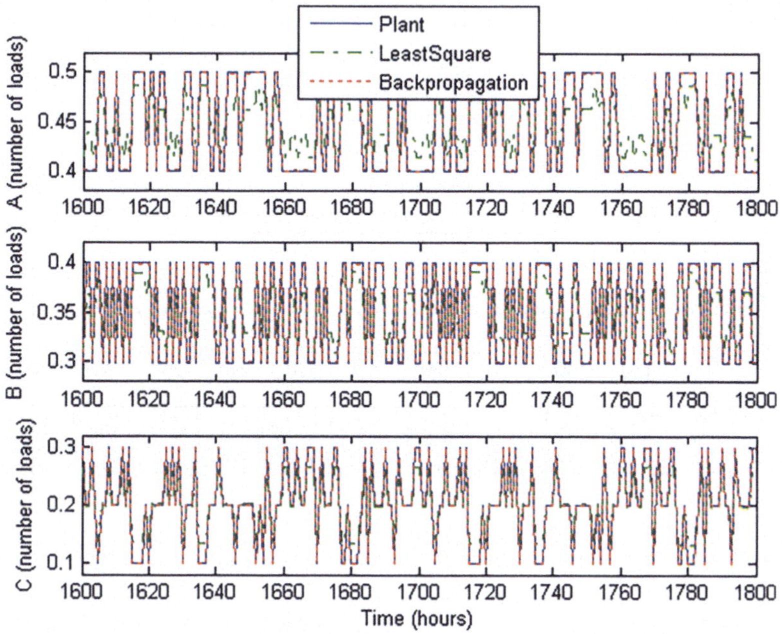

Fig. 2.6 Training results for Example 1

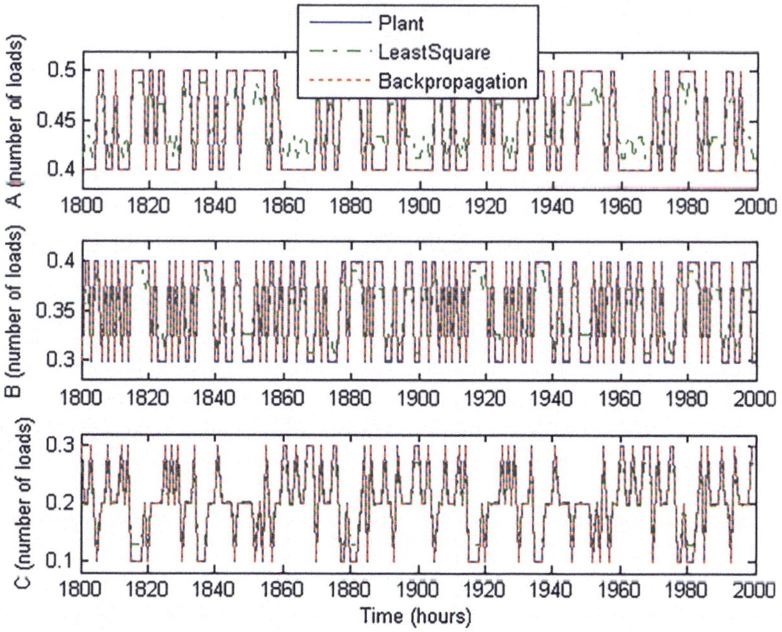

Fig. 2.7 Testing results for Example 1

Table 2.1 Results for Example 1

Methods	Training RMSE	Testing RMSE
Recursive least square	0.0717	0.0121
Backpropagation	0.0321	3.2561×10^{-5}
Sugeno fuzzy inference	*NAN*	

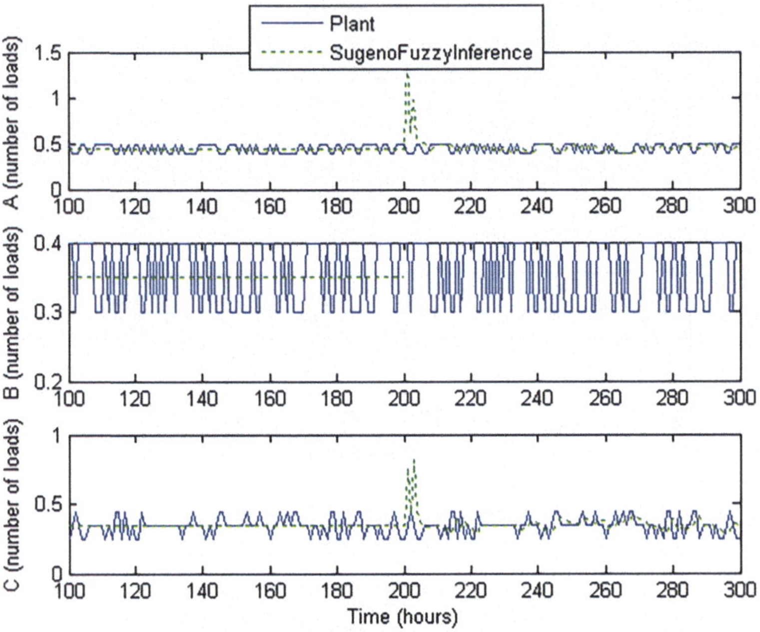

Fig. 2.8 Training for the Sugeno fuzzy inference system

From Table 2.1, it can be seen that the backpropagation algorithm achieves better accuracy when compared with the recursive least square because the training RMSE and the testing RMSE are smaller for the backpropagation algorithm. From Figs. 2.6 and 2.7, it can be seen that the backpropagation improves the recursive least square because the signal of the first one follows better the signal of the plant than the signal of the second one. From Fig. 2.8, the Sugeno fuzzy inference system is unstable for this prediction example, that is way it is important to guarantee the stability of the algorithms. Thus the backpropagation is good for the prediction problems.

Figure 2.9 shows the average of the identification error for the modified backpropagation algorithm. From this figure, it can be observed that the average of the identification error $\limsup_{T\to\infty} \frac{1}{T}\sum_{k=2}^{T} e_p^2(k)$ decreases, and it will converge to a value smaller than the upper bound of the uncertainty $\alpha_0\overline{\mu}^2$, as stated in Theorem 2.2.

The simulation of the weights' error for Theorem 2.3 cannot be obtained because the optimal weights which can minimize the modeling error are unknown [1].

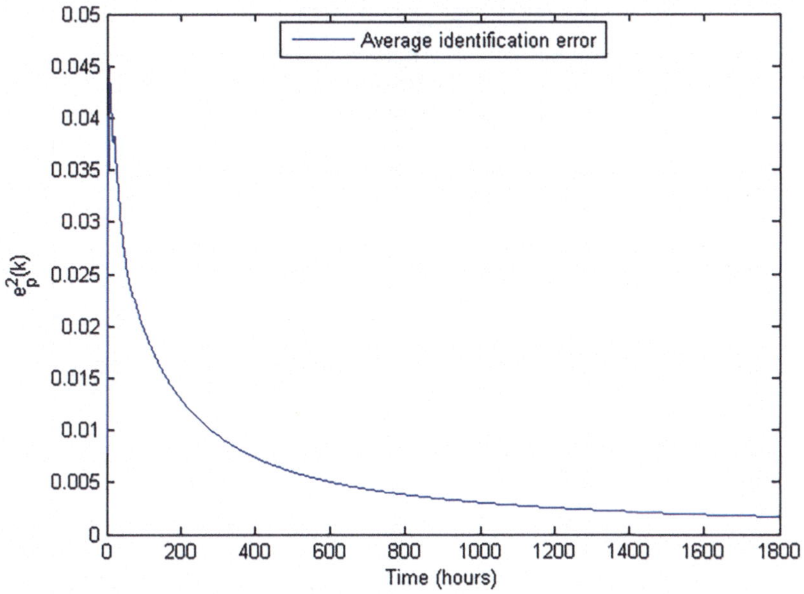

Fig. 2.9 Average identification error for Example 1

Example 2.2 Let us consider the nonlinear system given in an earlier study [31]:

$$y(k) = 0.3y(k-1) + 0.6y(k-2) + f(u(k-1)), \tag{2.19}$$

with $f(u(k-1)) = 0.6\sin(\pi u(k-1)) + 0.3\sin(3\pi u(k-1)) + 0.1\sin(5\pi u(k-1))$, and the input is $u(k-1) = \sin(8\pi(k-1)/200)$. In this example, the backpropagation algorithm given as (2.5), (2.12), and (2.15) is used for the identification of the nonlinear plant (2.19). The parameters of the backpropagation algorithm are $N = 2$, $M = 3$, V_{j1} and W_{ij1} are random numbers between 0 and 1, and $\alpha_0 = 0.25$. The backpropagation algorithm is compared with the recursive least square algorithm given by Goodwin and Sin [36] and used by Angelov and Filev [37] and Kasabov and Song [38] with parameters $P_1 = cI \in \Re^{2\times 2}$, where $c = 1$, V_1 is a random number between 0 and 1 and is compared with the Kalman filter algorithm given by Goodwin and Sin [36] and Haykin [25] and used by Rubio and Yu [2] with parameters $P_1 = cI \in \Re^{2\times 2}$, where $c = 1$, $R_1 = 0.1$, $R_2 = 1$, V_1 is a random number between 0 and 1. The training results are shown in Fig. 2.10, the testing results are shown in Fig. 2.11, and using (2.18) Table 2.2 shows the training and the testing RMSE results.

From Table 2.2, it can be seen that the backpropagation algorithm achieves better accuracy when compared with the recursive least square and the Kalman filter because the training RMSE and the testing RMSE are smaller for the backpropagation algorithm. From Figs. 2.10 and 2.11, it can be seen that the backpropagation

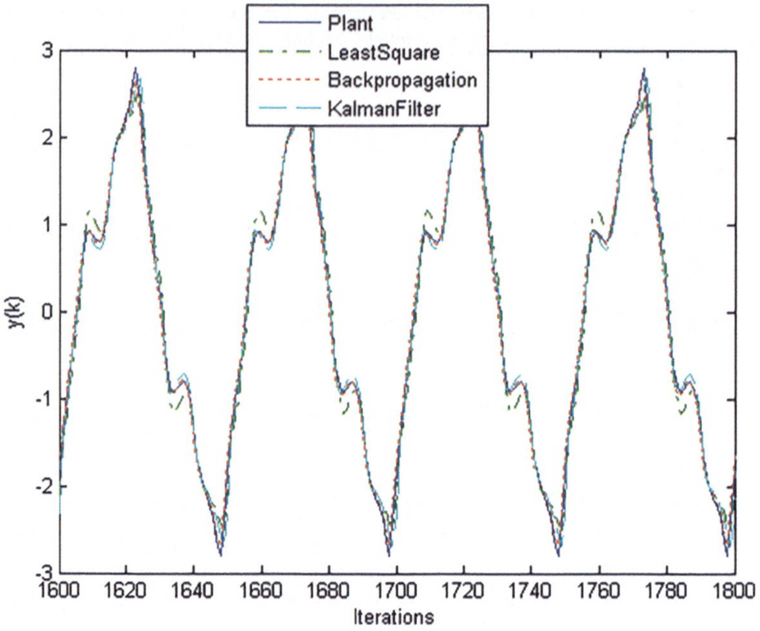

Fig. 2.10 Training results for Example 2

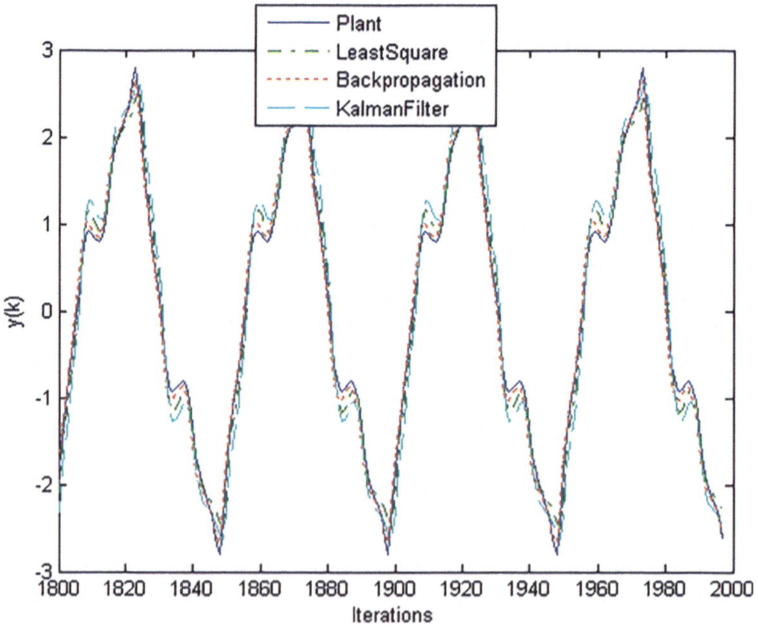

Fig. 2.11 Testing results for Example 2

Table 2.2 Results for Example 2

Methods	Training RMSE	Testing RMSE
Recursive least square	0.0714	0.0183
Kalman filter	0.0520	0.0283
Backpropagation	0.0413	0.0132

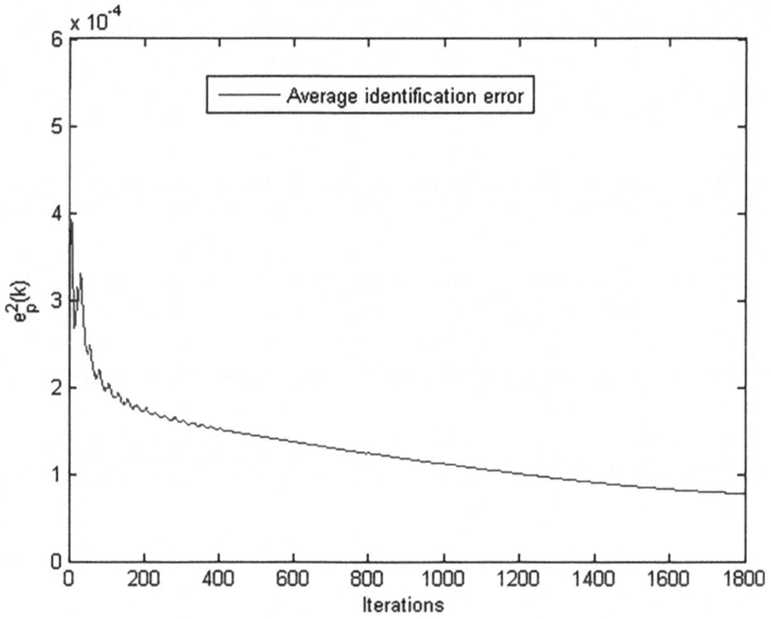

Fig. 2.12 Average identification error for Example 2

improves the recursive least square and the Kalman filter because the signal of the first follows better the signal of the plant than the signal of the second and the third. Thus, the backpropagation is good for the identification problems.

Figure 2.12 shows the average of the identification error for the modified backpropagation algorithm. From this figure, it can be observed that the average of the identification error $\limsup_{T\to\infty} \frac{1}{T}\sum_{k=2}^{T} e_p^2(k)$ decreases, and it will converge to a value smaller than the upper bound of the uncertainty $\alpha_0\overline{\mu}^2$, as stated in Theorem 2.2.

The simulation of the weights' error for Theorem 2.3 cannot be obtained because the optimal weights which can minimize the modeling error are unknown [1].

8 Concluding Remarks

In this chapter, it was proposed a theorem to assure the uniform stability of discrete-time systems, it was proven that the backpropagation algorithm with a new time varying rate is uniformly stable for online identification, the identification error converges to a small zone bounded by the uncertainty, and the weights' error are bounded by their initial weights' error. The backpropagation algorithm was compared with the recursive least square algorithm and with the Sugeno fuzzy inference system in the problem of the prediction of the distribution of loads each hour inside a warehouse, and the backpropagation algorithm was compared with the recursive least square and with the Kalman filter in an academic example. From the simulation results, it can be seen that the backpropagation algorithm achieved better accuracy when compared with the recursive least square algorithm and with the Kalman filter algorithm; in addition, the Sugeno fuzzy inference system was unstable. Thus, the backpropagation is good for the prediction and the identification problems. As a future work, a stable algorithm for the radial basis function will be proposed, a new algorithm for the feedforward neural network that guarantees asymptotic stability will be proposed, a method to find the number of neurons in the hidden layer online will be proposed, and the proposed algorithms will be applied for other real problems.

References

1. E.B. Kosmatopoulos, M.M. Polycarpou, M.A. Christodoulou, P.A. Ioannou, High-order neural network structures for identification of dynamic systems. IEEE Trans. Neural Networks **6**(2), 422–431 (1995)
2. J.J. Rubio, W. Yu, Nonlinear system identification with recurrent neural networks and dead-zone Kalman filter algorithm. Neurocomputing **70**, 2460–2466 (2007)
3. J.A.K. Suykens, J. Vandewalle, B.D. Moor, Lur'e systems with multilayer perceptron and recurrent neural networks: absolute stability and dissipativity. IEEE Trans. Autom. Control **44**(4), 770–774 (1999)
4. Z. Wang, H. Zhang, W. Yu, Robust exponential stability analysis of neural networks with multiple time delays. Neurocomputing **70**, 2534–2543 (2007)
5. Z. Wang, H. Zhang, W. Yu, Robust stability of Cohen-Grosberg neural networks via state transmission matrix. IEEE Trans. Neural Networks **20**(1), 169–174 (2009)
6. Z.B. Xu, H. Qiao, J. Peng, B. Zhang, A comparative study of two modeling approaches in neural networks. Neural Networks **17**, 73–85 (2004)
7. W. Yu, Passivity Analysis for dynamic multilayer neuro identifier. IEEE Trans. Circuits Syst. I Fund. Theory Appl. **50**(1), 173–178 (2003)
8. W. Yu, X. Li, Passivity analysis of dynamic neural networks, with different time-scales. Neural Proces. Lett. **25**, 143–155 (2007)
9. X. Yu, M. Onder, O. Kaynak, A general backpropagation algorithm for feedforward neural networks learning. IEEE Trans. Neural Networks **13**(1), 251–254 (2002)
10. J.J. Rubio, Stability Analysis for an online evolving neuro-fuzzy recurrent network, in *Evolving Intelligent Systems: Methodology and Applications* (John Wiley and Sons, Hoboken; IEEE Press, Piscataway, 2010), pp. 173–199. ISBN: 978-0-470-28719-4

11. E. Umez-Eronini, System Dynamics and Control (CL-Engineering, first edition, 1998). ISBN 0534944515
12. J.J. Rubio, W. Yu, Stability analysis of nonlinear system identification via delayed neural networks. IEEE Trans. Circuits Syst. II **54**(2), 161–165 (2007)
13. J.A.K. Suykens, B.D. Moor, J. Vandewalle, Robust local stability of multilayer recurrent neural networks. IEEE Trans. Neural Networks **11**(1), 222–229 (2000)
14. W. Yu, X. Li, Discrete-time neuro-identification without robust identification. IEE Proc. Control Theory Appl. **150**(3), 311–316 (2003)
15. W. Yu, X. Li, Fuzzy identification using fuzzy neural networks with stable learning algorithms. IEEE Trans. Fuzzy Syst. **12**(3), 411–420 (2004)
16. W. Yu, Nonlinear system identification, using discrete-time recurrent neural networks with stable learning algorithms. Inf. Sci. **158**, 131–147 (2004)
17. W. Yu, M.A. Moreno, F. Ortiz, System identification using hierarchical fuzzy neural networks with stable learning algorithm. J. Intell. Fuzzy Syst. **18**, 171–183 (2007)
18. C.I. Ezeife, Selecting and materializing horizontally partitioned warehouse views. Data Knowl. Eng. **36**, 185–210 (2001)
19. V. Giordano, J. Bing, D. Naso, F. Lewis, Integrated supervisory and operational control of a warehouse with a matrix-based approach. IEEE Trans. Autom. Sci. Eng. **5**(1), 53–70 (2008)
20. D.J. Berndt, A.R. Hevner, J. Studnicki, The catch data warehouse: support for community health care decision-making. Decis. Support Syst. **35**, 367–384 (2003)
21. H.R. Nemati, D.M. Steiger, L.S. Iyer, R.T. Herschel, Knowledge warehouse: an architectural integration of knowledge management, decision support, artificial intelligence and data warehousing. Decis. Support Syst. **33**, 143–161 (2002)
22. J. Xu, X. Yao, C.-H. Choi, G. Gou, Materialized view selection as constrained evolutionary optimization. IEEE Trans. Syst. Man Cybernet. Part C Appl. Rev. **33**(4), 458–467 (2003)
23. C.I. Byrnes, A. Isidori, J.C. Willems, Passivity, feedback equivalence, and the global stabilization of minimum phase nonlinear systems. IEEE Trans. Autom. Control **36**, 1228–1240 (1991)
24. J.J. Rubio, P. Angelov, J. Pacheco, Uniformly stable backpropagation algorithm to train a feedforward neural network. IEEE Trans. Neural Networks **22**(3), 356–366 (2011)
25. S. Haykin, *Neural Networks - A Comprehensive Foundation* (Macmillan College Publ. Co., New York, 1994)
26. J.R. Hilera, V.J. Martines, *Redes Neuronales Artificiales, Fundamentos, Modelos y Aplicaciones* (Addison Wesley Iberoamericana, Boston, 1995)
27. J.S.R. Jang, C.T. Sun, *Neuro-Fuzzy and Soft Computing* (Prentice Hall, Hoboken, 1996)
28. G. Cybenco, Approximation by superposition of sigmoidal activation function. Math. Control Signals Syst. **2**, 303–314 (1989)
29. J.J. Rubio, Sofmls, online self-organizing fuzzy modified least-squares network. IEEE Trans. Fuzzy Syst. **17**(6), 1296–1309 (2009)
30. C.F. Juang, C.T. Lin, An online self-constructing neural fuzzy inference network and its applications. IEEE Trans. Fuzzy Syst. **6**(1), 12–32 (1998)
31. L.X. Wang, *A Course in Fuzzy Systems and Control* (Prentice Hall, Englewood Cliffs, 1997)
32. S. Jagannathan, Control of a class of nonlinear discrete-time systems using multilayer neural networks. IEEE Trans. Neural Networks **12**(5), 1113–1120 (2001)
33. Y.H. Kim, F.L. Lewis, Optimal design of CMAC neural-network controller for robot manipulators. IEEE Trans. Syst. Man Cybernet. Part C Appl. Rev. **30**(1), 22–30 (2000)
34. C. Kwan, F.L. Lewis, D.M. Dawson, Robust neural network control of rigid-link electrically driven robots. IEEE Trans. Neural Networks **9**(8), 591–588 (1998)
35. G. Loreto, R. Garrido, Stable neurovisual servoing for robot manipulators. IEEE Trans. Neural Networks **17**(4), 953–965 (2006)
36. G.C. Goodwin, K.S. Sin, *Adaptive Filtering Prediction and Control* (Prentice Hall, Englewood Cliffs, 1984)
37. P.P. Angelov, D.P. Filev, An approach to online identification of Takagi-Sugeno fuzzy models. IEEE Trans. Syst. Man Cybernet. **32**(1), 484–498 (2004)

38. N.K. Kasabov, Q. Song, DENFIS: dynamic evolving neural-fuzzy inference system and its application for time-series prediction. IEEE Trans. Fuzzy Syst. **10**(2), 144–154 (2002)
39. N.K. Kasabov, Evolving fuzzy neural networks for supervised/unsupervised online knowledge-based learning. IEEE Trans. Syst. Man Cybernet. **31**(6), 902–918 (2001)

Chapter 3
Analytic Neural Network Model of a Wind Turbine

1 Introduction

Researchers are often trying to improve the total power of a wind turbine. The dynamic model of a wind turbine plays an important role in some applications as the control, classification, pattern recognition, or prediction.

There is some research about neural networks and fuzzy systems. In [1–7], the fuzzy systems are used as the structure of evolving fuzzy systems. In [8–13], the neural networks are used as the structure of evolving neural networks. New methods for exploring the evolution of social groups are mentioned in [14]. An approach to predict from a data stream of real estate sales transactions is presented in [15]. Considering the above studies, a multilayer neural network is a good alternative for the modeling of the wind turbine behavior.

There is some research about hybrid systems. In [16], the authors make the first attempt to develop a hybrid system by integrated case-based reasoning and artificial neural networks as a model for mobile phone company. The development of a multiscale hierarchical hybrid model based on finite element analysis and neural network computation to link mesoscopic scale and macroscopic is presented in [17] to simulate the process of bone remodeling. In [18], as an alternative method to analytical modeling approaches, this study uses the advantages of neural networks such as no required knowledge of internal system parameters, less computational effort, and a compact solution for multivariable problems. In [19], both analytical and soft computing approaches are used in predicting the performance of an indirect evaporative cooling. A hybrid model of a boiler developed with the application of

J. de J. Rubio, *Stability Analysis of Neural Networks and Evolving Intelligent Systems*, https://doi.org/10.1007/978-3-031-87282-2_3

both analytical modeling and artificial intelligence is described in [20]. Two models for estimating essential oil extraction yield from Anise, at high pressure condition, were used in [21]: mathematical modeling and artificial neural network modeling. Considering the aforementioned researches, a hybrid system with a multilayer neural network is a good alternative for the modeling of the wind turbine behavior.

Commonly, the intelligent algorithm is directly used for the modeling of the wind turbine behavior; however, in this chapter, the proposed analytic model provides a good approximation. Thus, a neural network is proposed as a compensator to improve the approximation obtained by the analytic model. The analytic neural network model is a hybrid system which learns the wind turbine behavior considering real data of the inputs, states, and output.

The chapter is structured as follows. In Sect. 2, the dynamic model of a windward wind turbine of three blades with a rotatory tower is mentioned. In Sect. 3, the proposed analytic neural network model is presented for the modeling of the wind turbine behavior. In Sect. 4, it is guaranteed that the error of the analytic neural network model is smaller than the error of the analytic model. In Sect. 5, the analytic model and analytic neural network model are compared for the modeling of two trajectories of the wind turbine behavior. Finally, in Sect. 6, the conclusion and future research are detailed.

2 Analytic Model of a Wind Turbine with a Rotatory Tower

The analytic model is described in this section as the first part of the proposed model. The following subsection describes the algorithm proposed by this study used to approximate the wind turbine behavior.

This model is divided into four parts: the first is the mechanic model, the second is the aerodynamic model, the third is the electric model, and finally, the fourth is the combination of the aforementioned models to obtain the final analytic model. The dynamic model of the wind turbine is, first, the equations that represent the change between the wind energy and mechanic energy, and second, the equations that represent the change between the mechanic energy and electric energy.

2.1 The Analytic Model

Define the state variables as $x_1 = i_2$, $x_2 = \theta_2$, $x_3 = \dot{\theta}_2$, $x_4 = i_1$, $x_5 = \theta_1$, $x_6 = \dot{\theta}_1$, the inputs as $u_1 = F_2$, $u_2 = V_1$, and the output as $y = V_2$. Consequently, the dynamic model is given as follows [22, 23]:

$$\begin{aligned}
\dot{x}_1 &= -\frac{(R_2+R_e)}{L_2}x_1 + \frac{k_2}{L_2}x_3 \\
\dot{x}_2 &= x_3 \\
\dot{x}_3 &= -\frac{k_{b2}}{m_2 l_{c2}^2}x_2 - \frac{b_{b2}}{m_2 l_{c2}^2}x_3 - \frac{2\pi k_{b2}}{3m_2 l_{c2}^2} + \frac{\cos(x_5)}{3m_2 l_{c2}^2}u_1 \\
\dot{x}_4 &= -\frac{R_1}{L_1}x_4 - \frac{k_1}{L_1}x_6 + \frac{1}{L_1}u_2 \\
\dot{x}_5 &= x_6 \\
\dot{x}_6 &= -\frac{3k_{b1}}{4.5m_2 l_{c2}^2}x_5 - \frac{3b_{b1}}{4.5m_1 l_{c2}^2}x_6 + \frac{k_m}{4.5m_2 l_{c2}^2}x_4 \\
y &= R_e x_1 \\
u_1 &= F_{2a} + F_{2b} + F_{2c} \\
F_{2a} &= \frac{1}{2x_3}\rho A C_p(\lambda,\beta)V_\omega^3 \\
F_{2b} &= \frac{1}{2x_3}\rho A C_p(\lambda,\beta)V_\omega^3 \\
F_{2c} &= \frac{1}{2x_3}\rho A C_p(\lambda,\beta)V_\omega^3 \\
\lambda &= \frac{x_3 R}{V_\omega},
\end{aligned} \tag{3.1}$$

where $C_p(\lambda,\beta) = c_1\left(\frac{c_2}{\lambda_i} - c_3\beta - c_4\right)e^{-c_5/\lambda_i} + c_6\lambda$, $\frac{1}{\lambda_i} = \frac{1}{\lambda+0.08\beta} - \frac{0.035}{\beta^3+1}$, θ_1 is the angular position of the tower motor in rad, θ_2 is the angular position of a wind turbine blade in rad, l_{c2} is the length of the wind turbine blade center in m, m_1 is the tower mass in kg, m_2 is the blade mass in kg, g is the acceleration gravity in m/s^2, l_1 is the constant length of the tower in m, l_{c1} is the length of the tower center in m, τ_{2a} is the torque of the generator moved by the blade in kgm^{2}rad/s^2, τ_{1a} is the torque of the motor used to move the tower in kgm^{2}rad/s^2, k_{b1} and k_{b2} are the spring effect presented when the blade is near to stop in kgm^2/s^2, b_{b1} and b_{b2} are the shock absorber in kgm^{2}rad/s, F_{2a}, F_{2b}, and F_{2c} are the force of the air received by the three blades. Equation (3.1) describes the assumption that the air goes in one direction, if $\theta_1 = 0$, then the maximum air intake moves the blades of the wind turbine, but if the tower turns to the left or to the right and θ_1 changes, then the wind turbine turns, and the air intake decreases; k_m is a motor magnetic flux constant of the tower in Wb, i_1 is the motor armature current of the tower in A, ρ is the air density in Kg/m^3, $A = \pi R^2$ is the area swept by the rotor blades in m^2 with radius R in m, V_ω is the wind speed in m/s, $C_p(\lambda,\beta)$ is the performance coefficient of the wind turbine, whose value is a function of the tip speed ratio λ, $c_1 = 0.5176$, $c_2 = 116$, $c_3 = 0.4$, $c_4 = 5$, $c_5 = 21$, and $c_6 = 0.0068$ are coefficients, β is the blade pitch angle in rad, k_1 is the motor back emf constant in Vs/rad, k_2 is the generator back emf constant in Vs/rad, R_1 is the motor armature resistance in Ω, R_2 is the generator armature resistance in Ω, L_1 is the motor armature inductance in H, L_2 is the generator armature inductance in H, V_1 is the motor armature voltage in V, V_2 is the generator armature voltage in V, and i_2 is the generator armature current in A. For the generator of this chapter $V_2 = R_e i_2$.

3 Analytic Neural Network Model of a Wind Turbine with a Rotatory Tower

The neural network is described in this section as the second part of the proposed model. The following subsection describes the algorithm proposed by this study used for the modeling of a wind turbine behavior.

Normally, the intelligent algorithm is directly used for the modeling of the wind turbine behavior; however, in this chapter, the analytic model of (3.1) yields a good approximation. The neural network of this chapter is used to improve the approximation obtained by the analytic model. The analytic neural network model learns the behavior considering real data of the inputs, states, and output, and the eight inputs for the intelligent algorithm are denoted as $z_1(k) = u_{1r}$, $z_2(k) = u_{2r}$, $z_3(k) = x_{1r}$, $z_4(k) = x_{2r}$, $z_5(k) = x_{3r}$, $z_6(k) = x_{4r}$, $z_7(k) = x_{5r}$, and $z_8(k) = x_{6r}$. The output of the analytic neural network model is $y_r(k) = y_r$, where r denotes the real data.

The following subsection describes the algorithm proposed by this study used for the modeling of the wind turbine behavior.

4 Analytic Neural Network Model

The stable backpropagation algorithm is developed with a new time varying rate to guarantee its uniformly stability for online identification and its identification error converges to a small zone bounded by the uncertainty. The weights' error is bounded by the initial weights' error, i.e., overfitting and local optimum are eliminated in the mentioned algorithm [12, 24].

Stable backpropagation algorithm is as follows [12, 24]:

(1) Obtain the output of the nonlinear system $y(k)$ with Eq. (3.1). Note that the nonlinear system may have the structure represented by Eq. (3.2); the parameter $n = 8$ is selected according to this nonlinear system.

$$y_r(k) = f\left[Z_k\right], \tag{3.2}$$

where $Z_k = [z_1(k) \ldots, z_i(k), \ldots, z_8(k)]^T \in \Re^{8\times 1}$ is the input vector, f is an unknown nonlinear function, $f \in C^\infty$, and $y_r(k)$ is the real data output of the wind turbine.

(2) Select the following parameters: V_1 and W_1 as random numbers between 0 and 1, m as an integer number, and α_0 as a positive value smaller than or equal to 1; obtain the output of the neural network $\widehat{y}(1)$ with Eq. (3.3). The analytic neural

network that approximates the real data output behavior of the wind turbine with rotatory tower $y_r(k)$ is as follows:

$$\begin{aligned}\widehat{o}_{NN}(k) &= y(k) + NN(k)\\ NN(k) &= V_k\Phi_k = \sum_{j=1}^{m} V_{jk}\phi_{jk}\\ \Phi_k &= \left[\phi_{1k},\ldots,\phi_{jk},\ldots,\phi_{mk}\right]^T\\ \phi_{jk} &= \tanh\left(\sum_{i=1}^{8} W_{ijk}z_i(k)\right),\end{aligned} \tag{3.3}$$

where $z_1(k)$, $z_2(k)$, $z_3(k)$, $z_4(k)$, $z_5(k)$, $z_6(k)$, $z_7(k)$, and $z_8(k)$ are the eight behavior inputs, and V_{jk+1} and W_{ijk+1} are the weights of the hidden and output layers, respectively. m is the neuron number in the hidden layer, and $y(k)$ is the analytic model output (3.1). ϕ_j is the hyperbolic tangent function.

(3) For each iteration k, obtain the output of the neural network $\widehat{o}_{NN}(k)$ with Eq. (3.3), also obtain the neural network error $e_{NN}(k)$ with Eq. (3.4), and update the parameters V_{jk+1} and W_{ijk+1} with Eq. (3.5).

$$e_{NN}(k) = \widehat{o}_{NN}(k) - y_r(k) \tag{3.4}$$

$$\begin{aligned}V_{jk+1} &= V_{jk} - \alpha_k\phi_{jk}e_{NN}(k)\\ W_{ijk+1} &= W_{ijk} - \alpha_k\sigma_{ijk}e_{NN}(k),\end{aligned} \tag{3.5}$$

where the new time varying rate α_k is

$$\alpha_k = \frac{\alpha_0}{2\left(\frac{1}{2} + \sum_{j=1}^{m}\phi_{jk}^2 + \sum_{j=1}^{m}\sum_{i=1}^{8}\sigma_{ijk}^2\right)},$$

where $i = 1,\ldots,8$, $j = 1,\ldots,m$, $\sigma_{ijk} = V_{jk}\mathrm{sech}^2(\sum_{i=1}^{N} W_{ijk}z_i(k))z_i(k) \in \Re$, α_0 is the constant learning speed, $\widehat{o}_{NN}(k)$ is the output of the analytic neural network model, and $y_r(k)$ is the real data output of the wind turbine.

Figure 3.1 shows the proposed analytic neural network model for the modeling of the wind turbine behavior.

Remark 3.1 There are two conditions for applying this algorithm for nonlinear systems: The first one is that the nonlinear system may have the form described by (3.1), and the second one is that the uncertainty $\mu(k)$ may be bounded.

Remark 3.2 The value of the parameter used for the stability of the algorithm $\overline{\mu}$ is unimportant, because this parameter is not used in the algorithm. The bound of

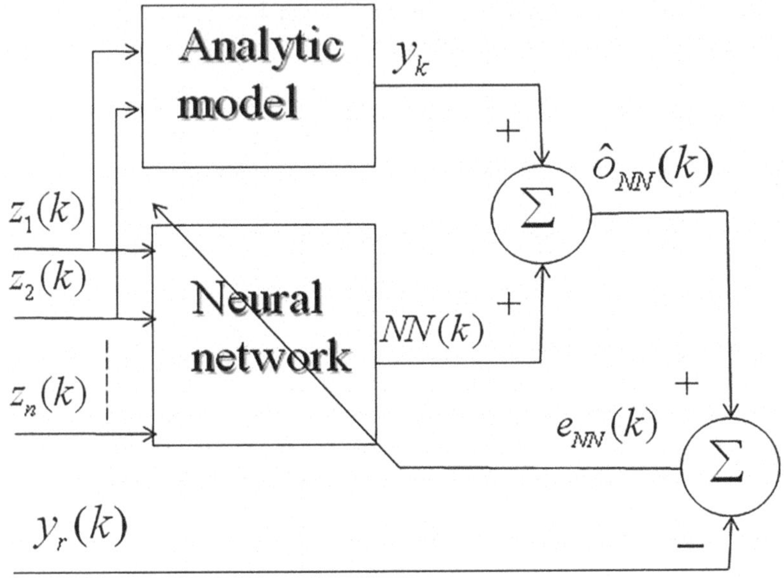

Fig. 3.1 Analytic neural network model

$\mu(k)$ is needed to guarantee the stability of the algorithm, but it is not used in the backpropagation algorithm (3.3), (3.4), (3.5).

Remark 3.3 The proposed analytic neural network has one hidden layer. It was reported in the literature that a feedforward neural network with one hidden layer is enough to approximate any nonlinear system.

Remark 3.4 Note that the behavior of the algorithm could be improved by changing the values of m or α_0.

5 Main Contribution of the Analytic Neural Network Model

In this section, the analytic neural network model will be guaranteed to be more approximated with the wind turbine behavior than the analytic model.

Define the analytic error as follows:

$$e(k) = y(k) - y_r(k), \tag{3.6}$$

where $y(k)$ is the output of the analytic model and $y_r(k)$ is the real data output of the wind turbine.

The real data output is rewritten as follows:

$$y_r(k) = y(k) + \sigma(k), \tag{3.7}$$

where $y(k)$ is the output of the analytic model, $y_r(k)$ is the real data output, and $\sigma(k)$ is the unmodeled dynamic.

Consider the following theorem.

Theorem 3.1 ([25]) *Suppose that the input universe of discourse U is a compact set in $\Re^n$. Then, for any given real continuous function $\sigma(k)$ on U and arbitrary $\in > 0$, there exists a neural network $NN(k)$ in the form (3.3) such that*

$$\sup_{x \in U} |NN(k) - \sigma(k)| < \in . \tag{3.8}$$

That is, the neural network $NN(k)$ is an approximator of $\sigma(k)$.

Proof *See [25] for the proof.*

The above theorem can be rewritten as follows:

Corollary 3.1 *The unmodeled dynamic $\sigma(k)$ of (3.7) is estimated by the neural network model $NN(k)$ of (3.3). It is written as follows:*

$$NN(k) \approx \sigma(k). \tag{3.9}$$

The following theorem shows the main contribution of this chapter.

Theorem 3.2 *The neural network error $e_{NN}(k)$ (3.4) of the analytic neural network model (3.3) for the modeling of the real data output of the wind turbine $y_r(k)$ is smaller than the analytic error (3.6) of the analytic model (3.1). It is written as follows:*

$$|e_{NN}(k)| \leq |e(k)| . \tag{3.10}$$

Proof *See [26] for the proof.*

Remark 3.5 The first difference of this analytic neural network model with the models considered by [1–12], and [13] is that the other models considered the neural network model as the unique algorithm for the approximation of the wind turbine behavior, while in this chapter the analytic neural network model (3.3) is the combination of two models: one is the analytic and the other is the neural network to obtain a better approximation. The second difference of analytic neural network model with the classical ones is that the second uses signum functions in the hidden and in the output layer being computational more complex, and the first uses a sigmoid function only in the hidden layer being more simple.

Remark 3.6 The difference of this analytic neural network model with the hybrid models considered by [16–20], and [21] is that the other studies are not applied for the modeling of the wind turbine behavior.

6 Experimental Results

The analytic neural network model of (3.3) is used for the modeling of the wind turbine output. The objective is that the analytic neural network model output $\widehat{o}_{NN}$ must be nearer with the real output of the wind turbine y_r than the analytic model output y. 8412 data are used for the training, and 2804 data are used for the testing. The root mean square error is used for the comparison results [10, 12, 13, 22]:

$$\text{RMSE} = \left(\frac{1}{N} \sum_{k=1}^{N} e_i^2(k) \right)^{\frac{1}{2}}, \tag{3.11}$$

where N is the number of iterations, and $e_i(k) = e(k)$ is the analytic error of (3.6), or $e_i(k) = e_{NN}(k)$ is the neural network error of (3.4).

Figure 3.2 shows the prototype of a wind turbine with a rotatory tower which is considered for the simulations of the analytic model. This prototype has three blades with a rotatory tower which does not use a gear box. Table 3.1 shows the parameters of the prototype. The parameters m_2 and l_{c2} are obtained from the wind

Fig. 3.2 Prototype of a wind turbine with rotatory tower

Table 3.1 Parameters of the prototype

Parameter	Value	Parameter	Value
l_{c2}	0.5 m	R_e	30Ω
m_2	0.5 kg	k_m	0.09 Wb
k_{b2}	1×10^{-6} kgm^2/s^2	k_{b1}	1×10^{-6} kgm^2/s^2
b_{b2}	1×10^{-1} kgm^2 rad/s	b_{b1}	1×10^{-1} kgm^2 rad/s
k_2	0.45 Vs/rad	k_1	0.0045 Vs/rad
R_2	6.96Ω	R_1	18Ω
L_2	6.031×10^{-1}H	L_1	6.031×10^{-1}H
R	l_{c2} m	V_ω	5 m/s
ρ	1.225 kg/m^3	β	0.5 rad
g	9.81 m/s^2		

turbine blades. The parameters R_1, L_1, and k_1 are obtained from the tower motor. The parameters k_2, R_2, R_e, and L_2 are obtained from the wind turbine generator. The parameters R, ρ, V_ω, and β are obtained from [27–29], and [30].

The dynamic model of the wind turbine with a rotatory tower is given by Eq. (3.1) with the parameters of Table 3.1. 1×10^{-5} is considered as the initial condition for the plant states $x_1 = i_2$, $x_2 = \theta_2$, $x_3 = \dot{\theta}_2$, $x_4 = i_1$, $x_5 = \theta_1$, and $x_6 = \dot{\theta}_1$.

6.1 Example 1

Example 1 considers the first movement of the wind turbine described as follows: (1) From 0 s to 2 s, both inputs are fed; consequently, the tower moves far from the maximum air intake, the generator current is decreased, and the wind turbine blades stop moving, (2) from 2 s to 4 s, both inputs are not fed; consequently, current is not generated, and both the tower and wind turbine blades do not move, (3) from 4 s to 6 s, both inputs are fed, but the air intake is positive and tower voltage is negative; consequently, the tower returns to the maximum air intake, the generator current is increased, and the wind turbine blades move, and (4) from 6 s to 8 s, both inputs are not fed; consequently, current is not generated, and the tower and wind turbine blades do not move.

The analytic model of (3.1) is used with parameters $x_1(1) = x_2(1) = x_3(1) = x_4(1) = x_5(1) = x_6(1) = 1 \times 10^{-5}$.

The analytic neural network model of (3.1), (3.3)–(3.5) is used with parameters $m = 4, \alpha_0 = 0.2$, $V_1 = $ rand, $W_1 = $ rand, rand is a random number, $x_1(1) = x_2(1) = x_3(1) = x_4(1) = x_5(1) = x_6(1) = 1 \times 10^{-5}$.

Figure 3.3 shows the modeling of the wind turbine behavior using the analytic model and analytic neural network model for the training. Figure 3.6 shows the modeling of the wind turbine behavior using the analytic model and analytic neural network model for the testing. RMSE for the analytic model and analytic neural network model is presented in Fig. 3.4 for the training and in Fig. 3.7 for the testing.

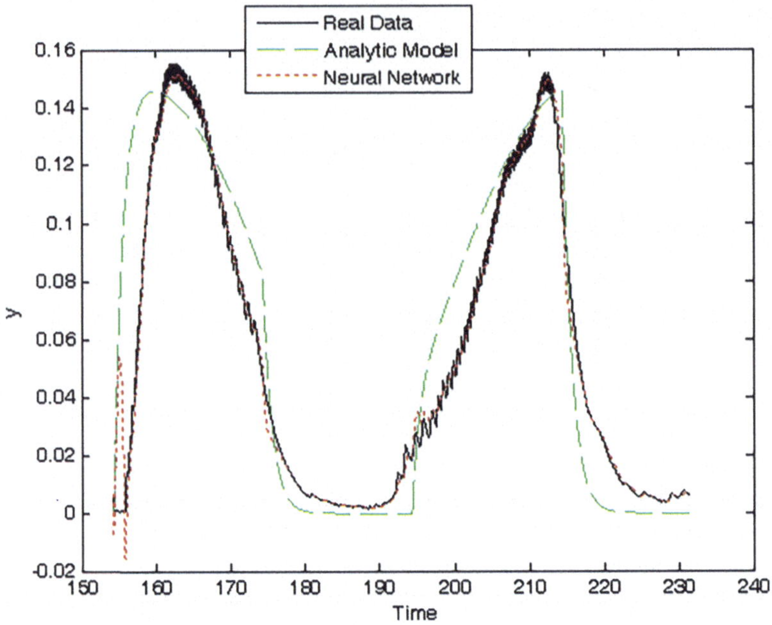

Fig. 3.3 Modeling for the training of Example 1

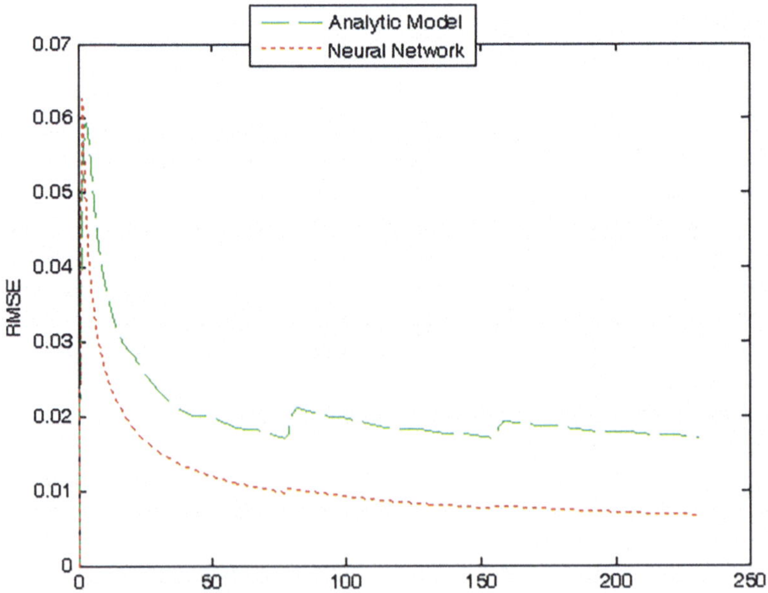

Fig. 3.4 RMSE for the training of Example 1

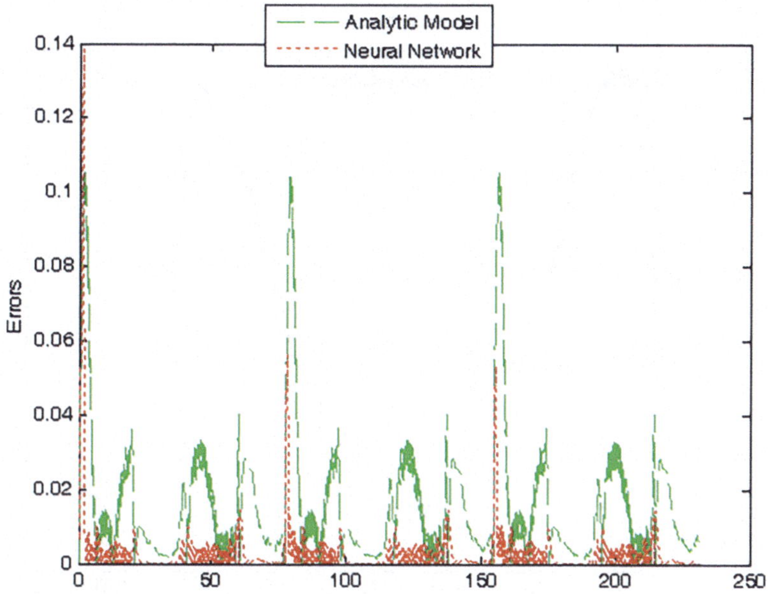

Fig. 3.5 Absolute error for the training of Example 1

Table 3.2 Comparison of the errors for Example 1

	RMSE for testing	RMSE for testing
Analytic model	0.0171	0.0086
Analytic neural network	0.0067	0.0067

Absolute errors of Theorem 3.2 for the analytic model and analytic neural network model are presented in Fig. 3.5 for the training and in Fig. 3.8 for the testing. Table 3.2 shows the root mean square error for the analytic model and analytic neural network model.

From Figs. 3.3, 3.4, and 3.5 and Table 3.2, it is shown that the analytic neural network model is the best for the training of the wind turbine behavior because the RMSE and absolute error of the above algorithm are the smallest ones. The training could be used for online designs as are the control or prediction.

From Figs. 3.6, 3.7, and 3.8 and Table 3.2, it is shown that the analytic neural network model is the best for the testing of the wind turbine behavior because the RMSE and absolute error of the above algorithm is the smallest one. The testing could be used for offline designs as are the pattern recognition or classification.

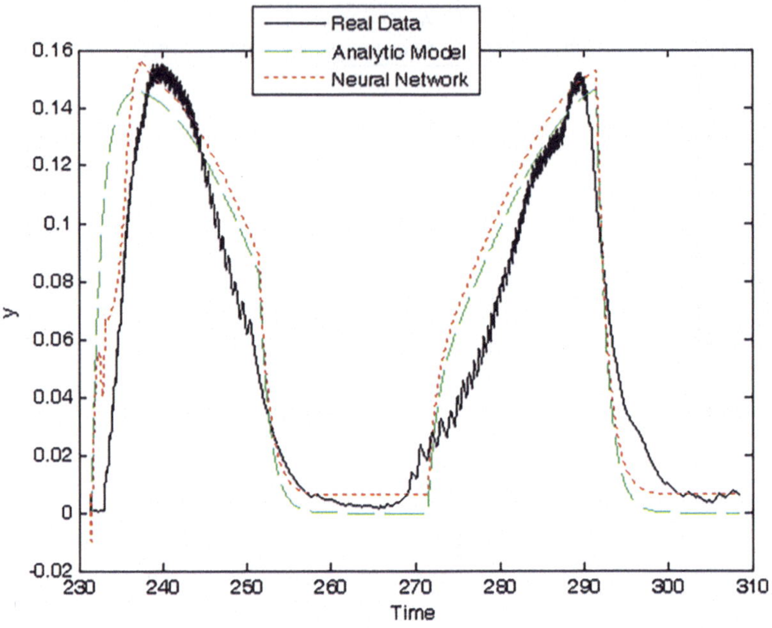

Fig. 3.6 Modeling for the testing of Example 1

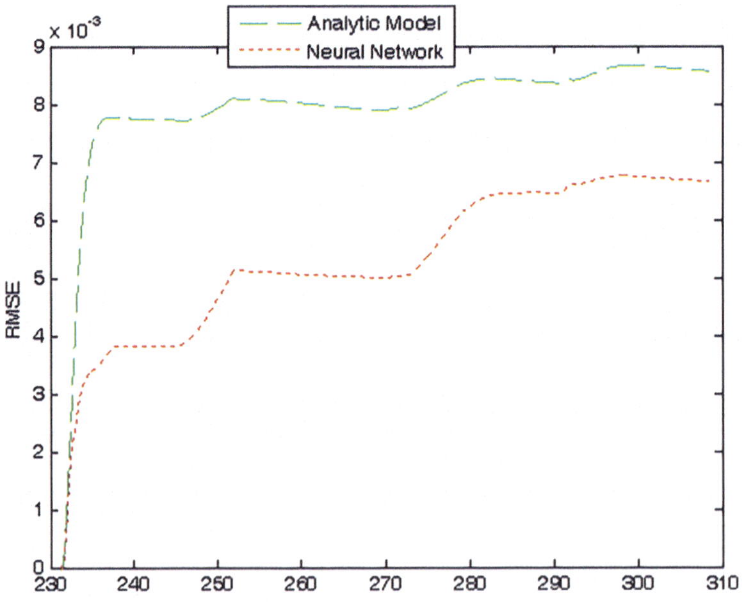

Fig. 3.7 RMSE for the testing of Example 1

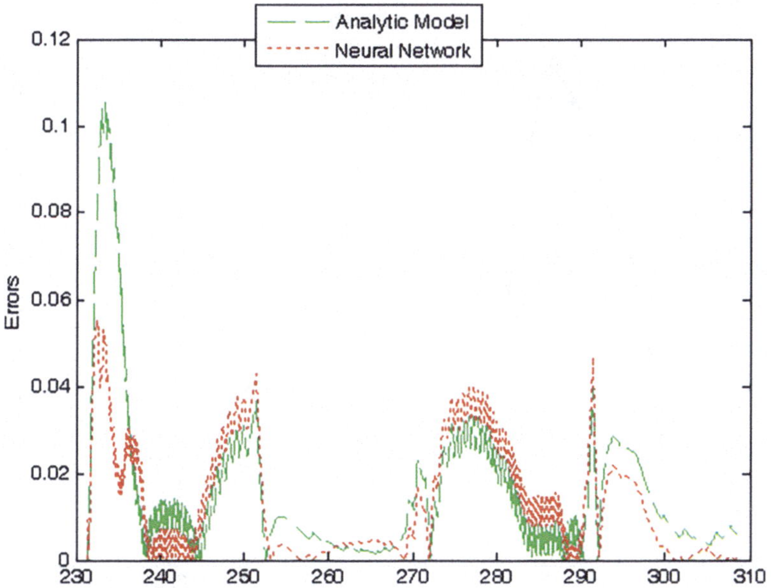

Fig. 3.8 Absolute error for the testing of Example 1

6.2 *Example 2*

Example 2 considers the second movement of the wind turbine described as follows: (1) From 0 s to 2 s, the input air is fed, and the tower input is not fed; consequently, the tower remains in the maximum air intake, the generator current is maximum, and the wind turbine blades have motion, (2) from 2 s to 4 s, the air is not fed, and the tower input is fed; consequently, current is not generated, the tower moves far from the maximum air intake, and the wind turbine blades do not have motion, (3) from 4 s to 6 s, the air is fed, and the tower input is not fed; consequently, the tower does not move, the generator current is minimum, and the wind turbine blades almost do not move; (4) from 6 s to 8 s, the air is not fed, and the tower input is fed with a negative voltage; consequently, current is not generated, the tower returns to the maximum air intake, and the wind turbine blades do not have motion.

The analytic model of (3.1) is used with parameters $x_1(1) = x_2(1) = x_3(1) = x_4(1) = x_5(1) = x_6(1) = 1 \times 10^{-5}$.

The analytic neural network model of (3.1), (3.3)–(3.5) is used with parameters $m = 4$, $\alpha_0 = 0.2$, $V_1 = \text{rand}$, $W_1 = \text{rand}$, rand is a random number, $x_1(1) = x_2(1) = x_3(1) = x_4(1) = x_5(1) = x_6(1) = 1 \times 10^{-5}$.

Figure 3.9 shows the modeling of the wind turbine behavior using the analytic model and analytic neural network model for the training. Figure 3.12 shows the modeling of the wind turbine behavior using the analytic model and analytic neural network model for the testing. RMSE for the analytic model and analytic neural network model is presented in Fig. 3.10 for the training and in Fig. 3.13 for the

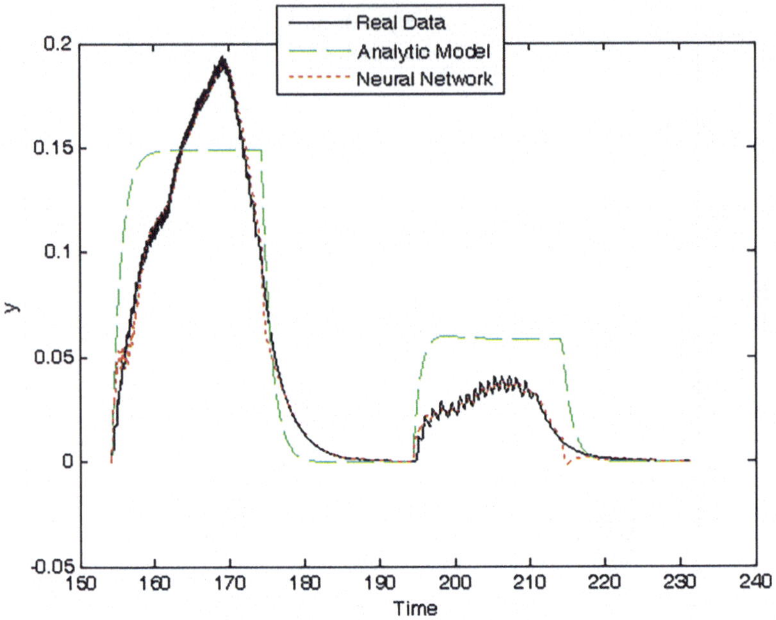

Fig. 3.9 Modeling for the training of Example 2

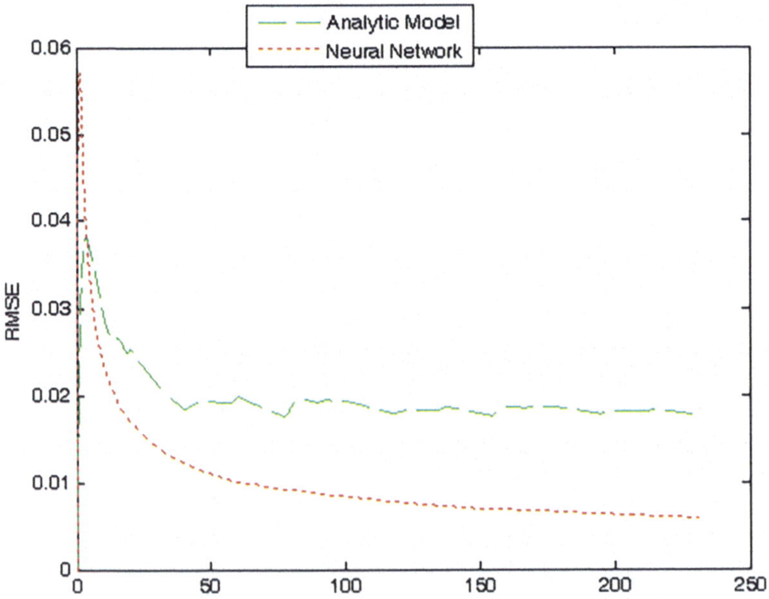

Fig. 3.10 RMSE for the training of Example 2

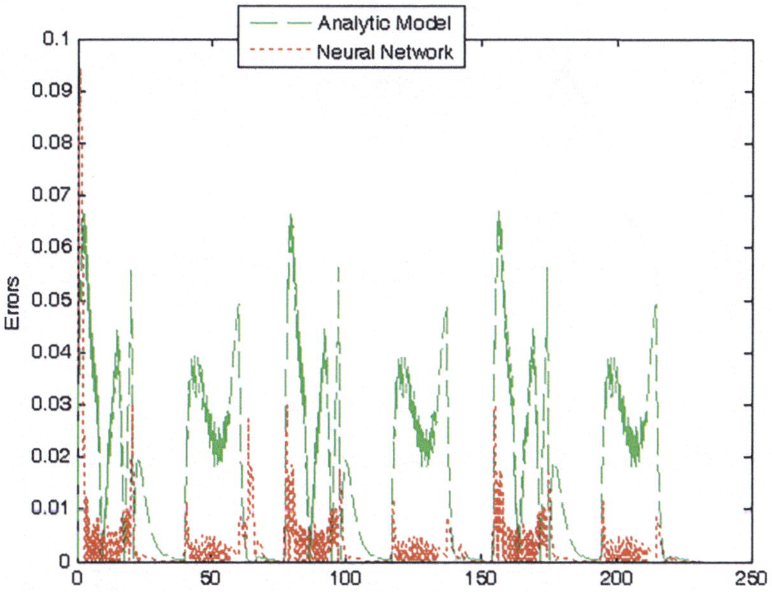

Fig. 3.11 Absolute error for the training of Example 2

Table 3.3 Comparison of the errors for Example 2

	RMSE for testing	RMSE for testing
Analytic model	0.0177	0.0089
Analytic neural network	0.0059	0.0072

testing. Absolute errors of the theorem 3.2 for the analytic model and analytic neural network model are presented in Fig. 3.11 for the training and in Fig. 3.14 for the testing. Table 3.3 shows the root mean square error for the analytic model and analytic neural network model.

From Figs. 3.9, 3.10, and 3.11 and Table 3.3, it is shown that the analytic neural network model is the best for the training of the wind turbine behavior because the RMSE and absolute error of the above algorithm is the smallest one. The training could be used for online designs as are the control or prediction.

From Figs. 3.12, 3.13, and 3.14 and Table 3.3, it is shown that the analytic neural network model is the best for the testing of the wind turbine behavior because the RMSE and absolute error of the above algorithm is the smallest one. The testing could be used for offline designs as are the pattern recognition or classification.

Remark 3.7 Choosing an appropriate number of neurons in the hidden layer is important in the behavior, because too many neurons result in a complex system that may be unnecessary for the problem and it can cause overfitting [12, 13, 31], whereas too few neurons produce a less powerful system that may be insufficient to

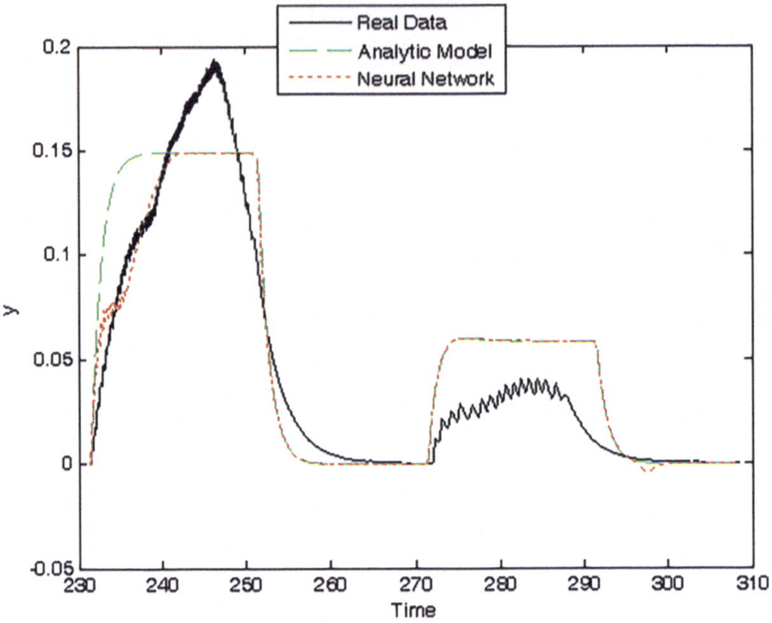

Fig. 3.12 Modeling for the testing of Example 2

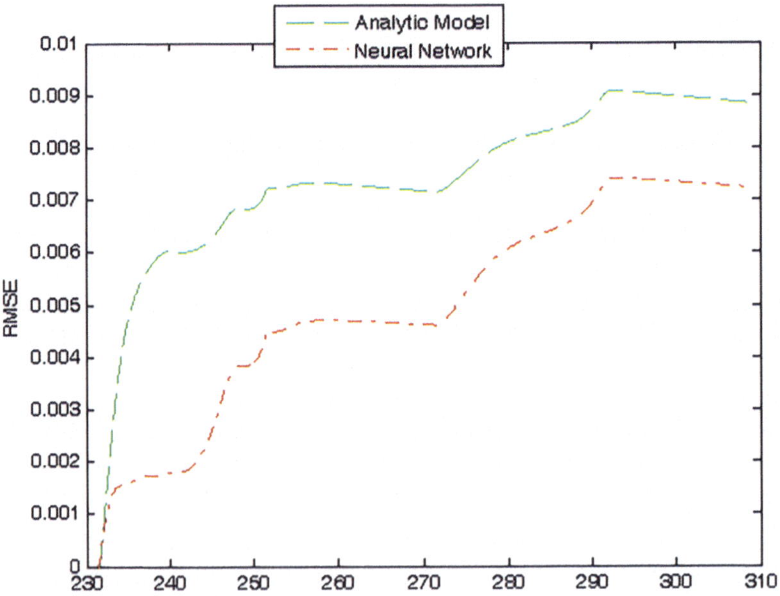

Fig. 3.13 RMSE for the testing of Example 2

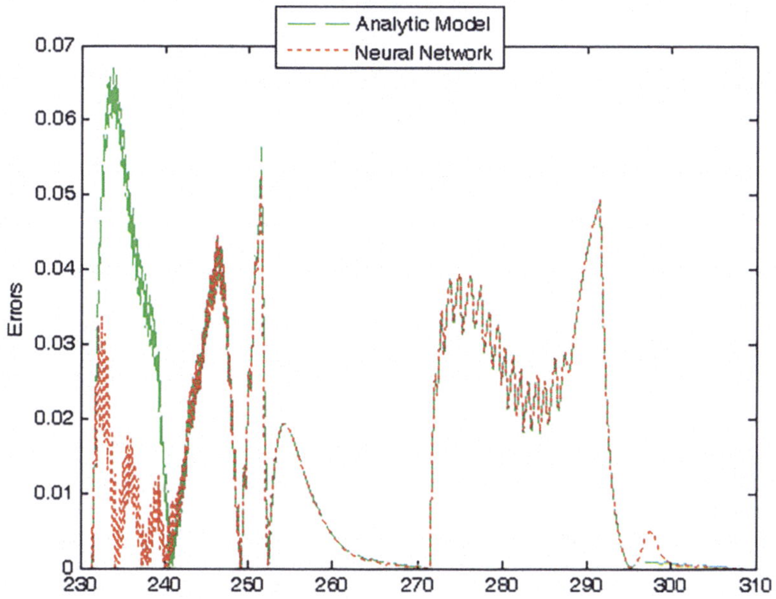

Fig. 3.14 Absolute error for the testing of Example 2

achieve the objective. The number of neurons is considered as a design parameter, and it is determined based on the trial-and-error method.

Remark 3.8 The difference between the two datasets considered in the experiments is that each dataset represents a different movement of the wind turbine [22].

7 Concluding Remarks

In this chapter, an analytic neural network model and an analytic model were compared for the modeling of the wind turbine behavior, giving that the analytic neural network model approach improved the analytic model, because the root mean square error for the first was the smallest one. The proposed technique could be used on control, prediction, pattern recognition, or classification. As a future research, the clustering algorithm will be used to estimate the number of hidden layer neurons, and the proposed modeling will be used in the design of interesting applications as are the control.

References

1. A. Buchachia, Dynamic clustering. Evol. Syst. **3**(3), 133–134 (2012)
2. E. Garcia-Cuesta, J.A. Iglesias, User modeling: through statistical analysis and subspace learning. Expert Syst. Appl. **39**(5), 5243–5250 (2012)
3. E. Lughofer, Single pass active learning with conflict and ignorance. Evol. Syst. **3**, 251–271 (2012)
4. E. Lughofer, B. Trawinski, K. Trawinski, O. Kempa, T. Lasota, On employing fuzzy modeling algorithms for the valuation of residential premises. Inf. Sci. **181**, 5123–5142 (2011)
5. E. Lughofer, *Evolving Fuzzy Systems - Methodologies, Advanced Concepts and Applications* (Springer, Berlin, 2011). ISBN: 978-3-642-18086-6
6. L. Maciel, A. Lemos, F. Gomide, R. Ballini, Evolving fuzzy systems for pricing fixed income options. Evol. Syst. **3**, 5–18 (2012)
7. M. Pratama, S.G. Anavatti, E. Lughofer, Genefis: towards an effective localist network. IEEE Trans. Fuzzy Syst. (2014). https://doi.org/10.1109/TFUZZ.2013.2264938
8. E. Balaguer-Ballester, H. Bouchachia, C.C. Lapish, Identifying sources of non-stationary neural ensemble dynamics. BMC Neurosci. **14**(Suppl 1), 15 (2013)
9. F. Bordignon, F. Gomide, Uninorm based evolving neural networks and approximation capabilities. Neurocomputing **127**, 13–20 (2014)
10. A. Marques Silva, W. Caminhas, A. Lemos, F. Gomide, A fast learning algorithm for evolving neo-fuzzy neuron. Appl. Soft Comput. **14**(B), 194–209 (2014)
11. M. Pratama, S.G. Anavatti, P.P. Angelov, E. Lughofer, PANFIS: a novel incremental learning machine. IEEE Trans. Neural Networks Learn. Syst. **25**(1), 55–68 (2014)
12. J.J. Rubio, Evolving intelligent algorithms for the modelling of brain and eye signals. Appl. Soft Comput. **14**(B), 259–268 (2014)
13. J.J. Rubio, J.H. Perez-Cruz, Evolving intelligent system for the modelling of nonlinear systems with dead-zone input. Appl. Soft Comput. **14**(B), 289–304 (2014)
14. P. Brodka, S. Saganowski, P. Kazienko, GED: the method for group evolution discovery in social networks, Soc. Netw. Anal. Min. **3**, 1–14 (2013)
15. B. Trawinski, Evolutionary fuzzy system ensemble approach to model real estate market based on data stream exploration. J. Univ. Comput. Sci. **19**(4), 539–562 (2013)
16. P.C. Chang, J.J. Lin, W.Y. Dzan, Forecasting of manufacturing cost in mobile phone products by case-based reasoning and artificial neural network models. J. Intell. Manuf. **23**, 517–531 (2012)
17. R. Hambli, H. Katerchi, C.L. Benhamou, Multiscale methodology for bone remodelling simulation using coupled finite element and neural network computation. Biomech. Model Mechanobiol. **10**, 133–145 (2011)
18. A. Naci-Celik, Artificial neural network modelling and experimental verification of the operating current of mono-crystalline photovoltaic modules. Sol. Energy **85**, 2507–2517 (2011)
19. T. Ravi-Kiran, S.P.S. Rajput, An effectiveness model for an indirect evaporative cooling (IEC) system: comparison of artificial neural networks (ANN), adaptive neuro-fuzzy inference system (ANFIS) and fuzzy inference system (FIS) approach. Appl. Soft Comput. **11**, 3525–3533 (2011)
20. H. Rusinowski, W. Stanek, Hybrid model of steam boiler. Energy **35**, 1107–1113 (2010)
21. A. Shokri, T. Hatami, M. Khamforoush, Near critical carbon dioxide extraction of anise (*Pimpinella anisum* l.) seed: mathematical and artificial neural network modeling, J. Supercrit. Fluids **58**, 49–57 (2011)
22. J.J. Rubio, L.A. Soriano, W. Yu, Dynamic model of a wind turbine for the electric energy generation. Math. Prob. Eng. **2014**, 1–8 (2014)
23. L.A. Soriano, W. Yu, J.J. Rubio, Modeling and control of wind turbine. Math. Prob. Eng. **2013**, 1–13 (2013)
24. J.J. Rubio, P. Angelov, J. Pacheco, An uniformly stable backpropagation algorithm to train a feedforward neural network. IEEE Trans. Neural Netw. **22**(3), 356–366 (2011)

25. L.X. Wang, *A Course in Fuzzy Systems and Control* (Pearson College Div, Facsimile edition, 1997). ISBN: 0-13-540882-2
26. J.J. Rubio, Analytic neural network model of a wind turbine. Soft Comput. **19**(12), 3455–3463 (2015)
27. E.B. Muhando, T. Senjyu, A. Yona, H. Kinjo, T. Funabashi, Disturbance rejection by dual pitch control and self-tuning regulator for wind turbine generator parametric uncertainty compensation. IET Control Theory Appl. **1**(5), 1431–1440 (2007)
28. C.Y. Tang, Y. Guo, J.N. Jiang, Nonlinear dual-mode control of variable-speed wind turbines with doubly fed induction generators. IEEE Trans. Control Syst. Technol. **19**(4), 744–756 (2011)
29. R. Vepa, Nonlinear optimal control of a wind turbine generator. IEEE Trans. Energy Convers. **26**(2), 468–478 (2011)
30. A. Zertek, G. Verbic, M. Pantos, Optimised control approach for frequency-control contribution of variable speed wind turbines. IET Renew. Power Gener. **6**(1), 17–23 (2012)
31. J.S.R. Jang, C.T. Sun, *Neuro-Fuzzy and Soft Computing* (Prentice Hall, Hoboken, 1996)

Chapter 4
Interpolation Neural Network Model of a Manufactured Wind Turbine

1 Introduction

The hybrid systems have been widely used in the learning of incomplete data for the applications of nonlinear modeling [1, 2], prediction [3], pattern recognition [4], classification [5, 6], control, fault detection and diagnosis in industrial systems [7, 8], visual inspection [9], and cascaded systems [10].

There are many studies about hybrid systems for the learning of nonlinear behaviors. Despite the proposals, few researches have been carried out in the past to perform the learning of incomplete data.

On the other hand, there are other methods for the learning of nonlinear behaviors with incomplete data, but they use noise signals considering the design as a stochastic problem, and it would be interesting to consider the design as a deterministic problem.

In this research, a hybrid algorithm as the combination of the stable neural network and interpolation algorithm is introduced for the learning of nonlinear systems with incomplete data where the design is considered as a deterministic problem. It consists in the following two stages.

First, the interpolation algorithm is used to obtain the missing data of all the variables in some nonlinear behavior. Figure 4.1 shows that the interpolation algorithm is applied to build the estimation of the variables denoted as $\widehat{x}_l(k)$ when only some points of the real variables denoted as $x_{lr}(k)$ are available.

Second, after the interpolation algorithm obtains the estimation of the variables, Fig. 4.2 shows that the interpolation neural network is employed to learn the output nonlinear behavior where the variables estimated by the interpolation algorithm denoted as $\widehat{x}_1(k) = \widehat{z}_1(k)$, $\widehat{x}_2(k) = \widehat{z}_2(k)$,..., $\widehat{x}_n(k) = \widehat{z}_n(k)$, $\widehat{x}_{n+1}(k) = \widehat{y}(k)$ are used instead of the real variables denoted as $x_{1,r}(k) = z_1(k)$, $x_{2,r}(k) = z_2(k)$, ..., $x_{n,r}(k) = z_n(k)$, $x_{n+1,r}(k) = y_r(k)$. $\widehat{y}(k)$ is the target output of the neural network. The inputs and output of the neural network are $\widehat{z}_1(k)$, $\widehat{z}_2(k)$,...,$\widehat{z}_n(k)$ and $NN(k)$, respectively. The importance of the neural network is that while the interpolation

J. de J. Rubio, *Stability Analysis of Neural Networks and Evolving Intelligent Systems*, https://doi.org/10.1007/978-3-031-87282-2_4

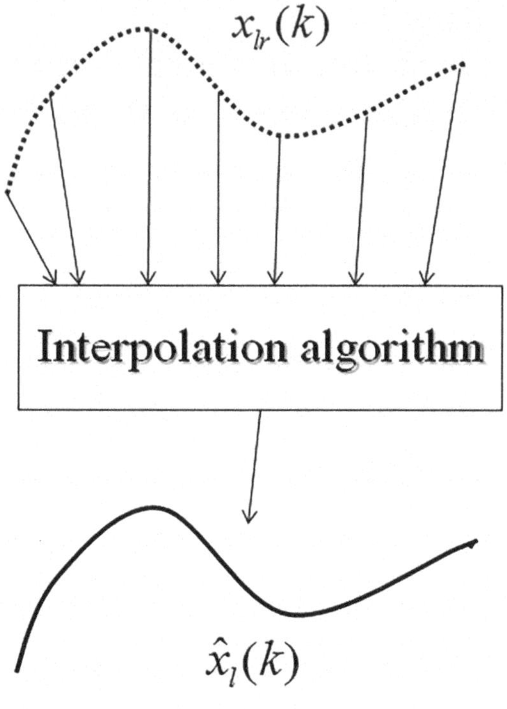

Fig. 4.1 Interpolation algorithm to estimate all the variables with incomplete data

algorithm only estimates the variables of the nonlinear behavior, the neural network learns the output behavior.

In the remainder of this section there will contain the survey of related works. Finally, the organization of this chapter will be mentioned.

1.1 Related Works

This subsection contains a survey of two kinds of related works: (a) hybrid systems for the learning of nonlinear behaviors and (b) methods for the learning of behaviors with incomplete data.

There is some research about the learning with hybrid systems. In [11], a learning approach to train uninorm-based hybrid neural networks is suggested. In [12], four semi-supervised learning methods are discussed. A specific ensemble strategy is developed in [13]. In [14], an approach to the construction of classifiers from imbalanced datasets is described. A dynamic pattern recognition method is proposed in [15]. In [16] and [17], the use of evolving classifiers for the activity recognition is described. Hybrid and ensemble methods in machine learning are focused in [18]. In [19], a granular neural network framework for the evolving fuzzy system modeling is introduced. A novel hybrid active learning strategy is proposed in [20].

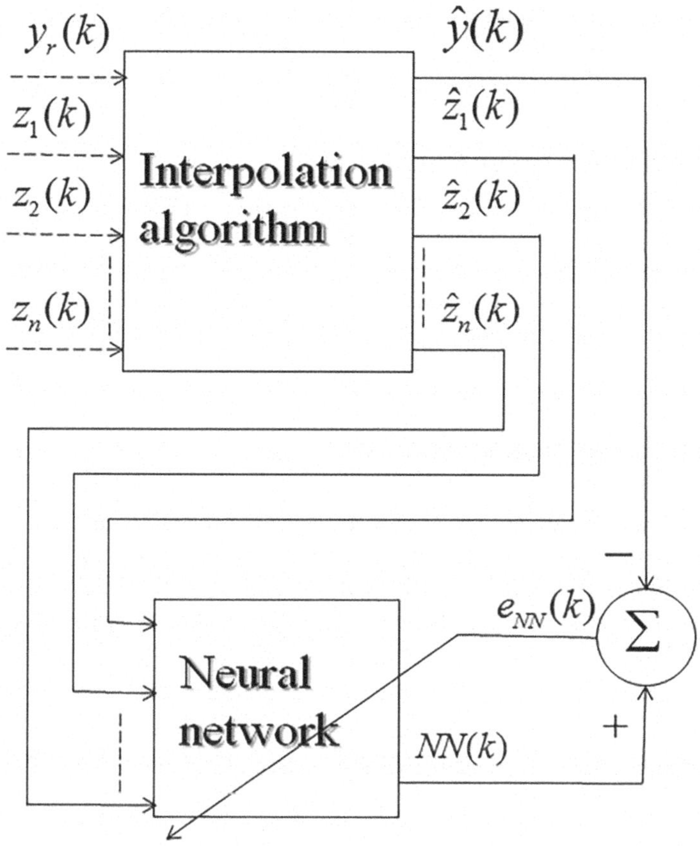

Fig. 4.2 Interpolation neural network for the learning

In [21], an enhanced version of the evolving participatory learning approach is developed. A class of hybrid-fuzzy models is designed in [22]. A parsimonious network based on the fuzzy inference system is addressed in [23]. In [1], a novel dynamic parsimonious fuzzy neural network is considered. A holistic concept of a fully data-driven modeling tool is proposed in [24]. In [5], a novel evolving fuzzy rule-based classifier is proposed. A novel meta-cognitive-based scaffolding classifier is considered in [25]. In [6], a novel interval type-2 fuzzy classifier is introduced. An evolving hybrid-fuzzy neural network-based modeling approach is introduced in [26].

Otherwise, there is some research about the learning of nonlinear behaviors with incomplete data. In [27], kernel regression method is used for the modeling with incomplete data. The story of incomplete and redundant representation modeling is introduced in [28]. In [29], the authors propose a new model called sparse hidden

Markov model. A novel sparse shape composition model is considered in [30]. In [31], a method is introduced for regression and classification problems.

1.2 Organization of the Chapter

The chapter is structured as follows. In Sect. 2, the interpolation neural network is described. In Sect. 3, the interpolation neural network is employed for the modeling of two trajectories of the wind turbine behavior. Finally, in Sect. 4, the conclusion and future research are detailed.

2 Interpolation Neural Network

This section is divided into two subsections which consider the two stages of the proposed algorithm. (a) The interpolation algorithm is utilized to estimate the nonlinear behavior of all the variables with incomplete data. (b) The interpolation neural network is employed for the learning of the nonlinear behavior output with incomplete data.

2.1 Interpolation Algorithm to Estimate the Incomplete Data

The interpolation algorithm is described in this subsection as the first part of the proposed model. The algorithm proposed in this part is used to estimate the missing data of all the variables with incomplete data, i.e., the proposed algorithm is a multidimension approximator where all the variables are independently estimated.

Description of the Interpolation Algorithm

Consider the functions $x_{lr}(k) = f(k_l) \in \Re$ with $l = 1, 2, , \ldots, n + 1$ is the number of variables estimated with this algorithm, $k_l = 1, 2, \ldots, T$, T are the iterations number for the variables, $x_{lr}(k)$ are the output real data of the nonlinear behaviors. The approximation consists in finding $\widehat{x_l}(k)$ such that they estimate the real variables with incomplete data $x_{lr}(k)$.

The slopes of $x_{lr}(k)$ denoted as $m_l(k)$ using the k_l and $x_{lr}(k)$ data of the nonlinear behavior are obtained as follows:

$$m_l(k) = \frac{x_{lr}(k) - x_{lr}(k-1)}{(k_l) - (k_l - 1)}. \tag{4.1}$$

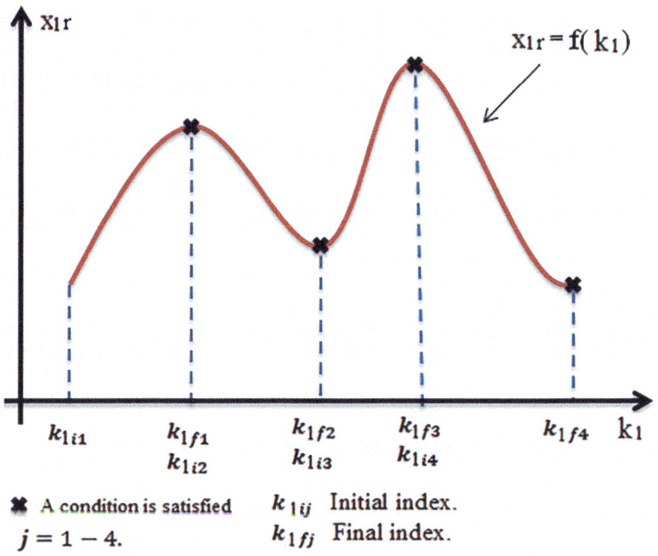

Fig. 4.3 Interpolation algorithm

The nonlinear behaviors are divided into N_l intervals, each interval is generated by considering the following inequality:

$$|(|m_l(k)| - |m_l(k-1)|)| \geq h_l, \quad (4.2)$$

where h_l is a small selected threshold parameter, and consider that the signals taken from k_l for each of the N_l intervals are denoted by j. Figure 4.3 shows the approximation of the nonlinear behaviors using the interpolation algorithm.

Equation (4.3) describes the approximation of the nonlinear behaviors using the proposed interpolation algorithm [32]:

$$\widehat{x_l}(k) = (1 - \lambda_l(k)) \cdot x_{l,i,j}(k) + \lambda_l(k) \cdot x_{l,f,j}(k), \quad (4.3)$$

where $x_{l,i,j}(k)$ are the initial values of $x_{lr}(k)$ in the interval j, $x_{l,f,j}(k)$ are the final values of $x_{lr}(k)$ in the interval j, k_l are the variant iterations inside of the interval j, $\lambda_l(k)$ are the variant-in-time parameters of the interval j, and $\lambda_l(k)$ are given as follows:

$$\lambda_l(k) = \frac{k_l - k_{l,i,j}}{k_{l,f,j} - k_{l,i,j}}, \quad (4.4)$$

where $k_{l,i,j}$ are the initial values of $\lambda_l(k)$ in the interval j, and $k_{l,f,j}$ are the final values of $\lambda_l(k)$ in the interval j.

It is known that $k_{l,i,j} \leq k_l \leq k_{l,f,j}$ for each interval j; consequently, $0 \leq \lambda_l(k) \leq 1$, and $\lambda_l(k)$ always increases. The variant parameters $\lambda_l(k)$ are important in the proposed interpolation algorithm because $\widehat{x_l}(k)$ are the approximations of $x_{lr}(k)$ from the initial points to the final points for each interval j. The interpolation algorithm for the approximation of nonlinear behaviors is as follows:

(1) Obtain the slope of $x_{lr}(k)$ denoted as $m_l(k)$ using the k_l and $x_{lr}(k)$ data of the nonlinear behaviors using Eq. (4.1), and select the threshold parameters h_l.
(2) Obtain the elements' number in the intervals N_l with Eq. (4.2).
(3) The intervals are denoted by j.
(4) For each interval j, obtain $\lambda_l(k)$ with Eq. (4.4).
(5) For each interval j, obtain $\widehat{x_l}(k)$ as the approximations of $x_{lr}(k)$ using Eq. (4.3).

Boundedness of the Interpolation Algorithm

In this section the variables of the interpolation algorithm will be guaranteed to be bounded. Substituting (4.4) into (4.3) of the interpolation algorithm gives

$$\widehat{x_l}(k) = \left(1 - \frac{k_l - k_{l,i,j}}{k_{l,f,j} - k_{l,i,j}}\right) \cdot x_{l,i,j}(k) + \frac{k_l - k_{l,i,j}}{k_{l,f,j} - k_{l,i,j}} \cdot x_{l,f,j}(k). \tag{4.5}$$

Equation (4.5) can be rewritten as follows:

$$\widehat{x_l}(k) = x_{l,i,j}(k) + \frac{k_l - k_{l,i,j}}{k_{l,f,j} - k_{l,i,j}} \left(x_{l,f,j}(k) - x_{l,i,j}(k)\right). \tag{4.6}$$

Theorem 4.1 *The outputs $\widehat{x_l}(k)$ of the interpolation algorithm (4.3)–(4.4), (4.6) are guaranteed to be bounded by $x_{l,i,j}(k)$ and by $x_{l,f,j}(k)$ for all the intervals j.*

Proof *See [33] for the proof.*

Remark 4.1 There are three differences between the interpolation algorithm introduced by Rubio et al. [32] and that considered in this study. The first difference is that in [32], the interval number is obtained by the changes in the slopes sign, while in this study the interval number is determined by Eq. (4.2). The second difference is that in [32], the interpolation algorithm is applied only to estimate the nonlinear system output, while in this chapter the interpolation algorithm is used to estimate all the nonlinear system variables. The third difference is that in [32], the interpolation algorithm is considered alone, while in this research the interpolation algorithm is combined with a stable neural network.

2.2 Neural Network to Learn with Incomplete Data

The neural network is described in this subsection as the second part of the proposed model. This subsection describes the algorithm proposed in this study for the modeling of a nonlinear behavior with incomplete data.

Description of the Neural Network

In this study, incomplete data are considered; consequently, the neural network of this chapter is used to learn the nonlinear behavior using only the variables estimated with the interpolation model, not the real data variables, that is, the variables of the interpolation algorithm $\widehat{z}_1(k)$, $\widehat{z}_2(k)$,...,$\widehat{z}_n(k)$, $\widehat{y}(k)$ are used instead of the real variables with incomplete data $z_1(k)$, $z_2(k)$, ..., $z_n(k)$, $y_r(k)$.

The stable backpropagation algorithm is employed with a new time varying rate to guarantee its uniformly stability for online identification and its identification error converge to a small zone bounded by the uncertainty. The weights' error is bounded by the initial weights' error, i.e., overfitting and local optimum are eliminated in the mentioned algorithm [2, 3].

Stable backpropagation algorithm is as follows [2, 3]:

(1) Obtain the output of the nonlinear system $y(k)$. Note that the nonlinear system may have the structure represented by Eq. (4.7); the parameter n is selected according to this nonlinear system.

$$\widehat{y}(k) = f\left[Z(k)\right], \tag{4.7}$$

where $Z(k) = [\widehat{z}_1(k)\ldots,\widehat{z}_i(k),\ldots,\widehat{z}_n(k)]^T \in \Re^{n\times 1}$ is the input vector, f is an unknown nonlinear function, $f \in C^\infty$, and $\widehat{y}(k)$, $\widehat{z}_1(k)$, $\widehat{z}_2(k)$,...,$\widehat{z}_n(k)$ are the outputs of the interpolation algorithm.

(2) Select the following parameters: $V(1)$ and $W(1)$ as random numbers between 0 and 1, m as an integer number, and α_0 as a positive value smaller than or equal to 1; obtain the output of the neural network $NN(1)$ with Eq. (4.8). The interpolation neural network that learns the real output with incomplete data of the nonlinear behavior $y_r(k)$ is as follows:

$$\begin{aligned}
NN(k) &= V(k)\Phi(k) = \sum_{j=1}^{m} V_j(k)\phi_j(k) \\
\Phi_k &= \left[\phi_1(k),\ldots,\phi_j(k),\ldots,\phi_m(k)\right]^T \\
\phi_j(k) &= \tanh\Big(\sum_{i=1}^{n} W_{ij}(k)\widehat{z}_i(k)\Big),
\end{aligned} \tag{4.8}$$

where $\widehat{z}_1(k)$, $\widehat{z}_2(k)$,...,$\widehat{z}_n(k)$ are the input estimation with the interpolation algorithm, and $V_j(k+1)$ and $W_{ij}(k+1)$ are the weights of the hidden and output layers, respectively. m is the neuron number in the hidden layer. ϕ_j is the hyperbolic tangent function.

(3) For each iteration k, obtain the output of the neural network $NN(k)$ with Eq. (4.8), also obtain the neural network error $e_{NN}(k)$ with Eq. (4.9), and update the parameters $V_j(k+1)$ and $W_{ij}(k+1)$ with Eq. (4.10).

$$e_{NN}(k) = NN(k) - \widehat{y}(k) \tag{4.9}$$

$$\begin{aligned} V_j(k+1) &= V_j(k) - \alpha(k)\phi_j(k)e_{NN}(k) \\ W_{ij}(k+1) &= W_{ij}(k) - \alpha(k)\sigma_{ij}(k)e_{NN}(k), \end{aligned} \tag{4.10}$$

where the new time varying rate $\alpha(k)$ is

$$\alpha(k) = \frac{\alpha_0}{2\left(\frac{1}{2} + \sum_{j=1}^{m}\phi_j^2(k) + \sum_{j=1}^{m}\sum_{i=1}^{n}\sigma_{ij}^2(k)\right)},$$

where $i = 1, \ldots, n$, $j = 1, \ldots, m$, $\sigma_{ij}(k) = V_j(k)\text{sech}^2(\sum_{i=1}^{n} W_{ij}(k)z_i(k))\widehat{z}_i(k) \in \Re$, α_0 is the constant learning speed, $\widehat{y}(k)$ is the output estimation with the interpolation algorithm, $NN(k)$ is the output of the interpolation neural network, and $\widehat{z}_1(k)$, $\widehat{z}_2(k)$,...,$\widehat{z}_n(k)$, $\widehat{y}(k)$ are the outputs of the interpolation algorithm.

Remark 4.2 The hyperbolic tangent is used as the activation function in the proposed neural network because it considers positive and negative values, being it more complete than others as the sigmoid function which only considers positive values.

Stability Analysis of the Neural Network

The following theorem guarantees that the interpolation neural network can approximate a nonlinear behavior.

Theorem 4.2 ([34]) *Suppose that the input universe of discourse U is a compact set in $\Re^n$. Then, for any given real continuous function $\sigma(k)$ on U and arbitrary $\in > 0$, there exists an interpolation neural network $NN(k)$ in the form (4.8) such that*

$$\sup_{x \in U} |NN(k) - \widehat{y}(k)| < \in . \tag{4.11}$$

That is, the neural network $NN(k)$ is an approximator of the output of the interpolation algorithm $\widehat{y}(k)$.

Proof *See [34] for the proof.*

The following theorem gives the stability of the neural network model.

Theorem 4.3 *The interpolation neural network (4.8), (4.9), and (4.10) applied for the identification of the nonlinear system (4.7) is uniformly stable, and the upper bound of the average identification error $e_p^2(k)$ satisfies*

$$\limsup_{T\to\infty}\frac{1}{T}\sum_{k=2}^{T}e_p^2(k)\le\alpha_0\overline{\mu}^2, \tag{4.12}$$

where $e_p^2(k)=\frac{\alpha(k-1)}{2}e^2(k-1)$, $0<\alpha_0\le 1\in\Re$ and $0<\alpha(k)\in\Re$ are defined in (4.10), $e(k)$ is defined in (4.9), $\mu(k)=y(k)-\sum_{j=1}^{M}V_j^*\phi_j^*$ is an uncertainty, $\overline{\mu}$ is the upper bound of the uncertainty $\mu(k)$, $|\mu(k)|<\overline{\mu}$, $\phi_j^*=\tanh(\sum_{i=1}^{N}W_{ij}^*x_i(k))$, and V_j^* and W_{ij}^* are unknown weights such that the uncertainty $\mu(k)$ is minimized.

Proof *See [2, 3] for the proof.*

The following theorem proves that the weights of the interpolation neural network are bounded.

Theorem 4.4 *When the average error $e_p^2(k)$ is bigger than the uncertainty $\alpha_0\overline{\mu}^2$, the weights' error is bounded by the initial weights' error as follows:*

$$\begin{aligned}&e_p^2(k+1)\ge\alpha_0\overline{\mu}^2\\ &\Longrightarrow\sum_{j=1}^{M}\widetilde{V}_j^2(k+1)+\sum_{j=1}^{M}\sum_{i=1}^{N}\widetilde{W}_{ij}^2(k+1)\le\sum_{j=1}^{M}\widetilde{V}_j^2(1)+\sum_{j=1}^{M}\sum_{i=1}^{N}\widetilde{W}_{ij}^2(1),\end{aligned} \tag{4.13}$$

where $i=1,\ldots,N$, $j=1,\ldots,M$, $\widetilde{V}_j(k)$ and $\widetilde{W}_{ij}(k)$ are the weights' error, $\widetilde{V}_j(1)$ and $\widetilde{W}_{ij}(1)$ are the initial weights' error, $e_p^2(k+1)=\frac{\alpha(k)}{2}e^2(k)$, $V_j(k+1)$, $W_{ij}(k+1)$, $0<\alpha_0\le 1\in\Re$, and $0<\alpha(k)\in\Re$ are defined in (4.10), $e(k)$ is defined in (4.9), $\overline{\mu}$ is the upper bound of the uncertainty $\mu(k)$, $|\mu(k)|<\overline{\mu}$.

Proof *See [2, 3] for the proof.*

Remark 4.3 There are two conditions for applying this algorithm for nonlinear systems: The first one is that the nonlinear system may have the form described by (4.7), and the second one is that the uncertainty $\mu(k)$ may be bounded.

Remark 4.4 The value of the parameter $\overline{\mu}$ used for the stability of the algorithm is unimportant, because this parameter is not used in the algorithm. The bound of $\mu(k)$ is needed to guarantee the stability of the algorithm, but it is not used in the backpropagation algorithm (4.8), (4.9), (4.10).

Remark 4.5 There is one important difference between the stable neural network of [2, 3] and the one considered in this study. It is that in [2, 3], the stable neural network is alone used for the learning of short data, while, in this research, the stable neural network is combined with the interpolation algorithm for the learning of nonlinear systems with incomplete data.

Remark 4.6 The fuzzy slopes model of [35] has two differences with the interpolation neural network of this research: (1) The fuzzy slopes model uses a fuzzy inference system, while the interpolation neural network employs the stable neural network, obtaining an advantage in the proposed method because a stable algorithm guarantees that all the variables will remain bounded, and (2) the fuzzy slopes model only considers the output with incomplete data, while the interpolation neural network considers all the variables with incomplete data, obtaining an advantage in the introduced technique because it is a generalization of the previous one.

3 Experimental Results

The interpolation neural network is compared with the fuzzy slopes model of [35] for the learning of the wind turbine behavior with incomplete data. The objective is that the interpolation neural network output NN of (4.1)–(4.4), (4.8)–(4.10) must be nearer with the real output of the wind turbine y_r than the fuzzy slopes model output.

Figure 4.4 shows the prototype of the manufactured wind turbine with a rotatory tower which is considered for this study. This prototype has three blades with a rotatory tower which does not use a gear box. Important research about wind turbines is presented in [4, 7, 36]. Table 4.1 shows the parameters of the prototype. The parameters m_2 and l_{c2} are obtained from the wind turbine blades. The parameters R_1, L_1, and k_1 are obtained from the tower motor. The parameters k_2, R_2, R_e, and L_2 are obtained from the wind turbine generator. The parameters R, ρ, V_ω, and β are obtained from [36].

1×10^{-5} is considered as the initial condition for the plant states $x_1 = i_2, x_2 = \dot{\theta}_2$, $x_3 = \theta_2$, $x_4 = i_1$, $x_5 = \dot{\theta}_1$, and $x_6 = \theta_1$. u_1 is the force of the air received by the three blades in km^{2}rad/s^2, u_2 is the motor armature voltage in V, θ_1 is the angular position of the tower motor in rad, θ_2 is the angular position of a wind turbine blade in rad, i_1 is the motor armature current of the tower in A, i_2 is the generator armature current in A, and y is the output voltage generated by the wind turbine in V. An electronic circuit and a microcontroller board of Arduino are used to digitalize and to send the obtained signals to a personal computer. Figure 4.5 shows the real

Fig. 4.4 Prototype of the manufactured wind turbine

Table 4.1 Parameters of the prototype

Parameter	Value	Parameter	Value
l_{c2}	0.5 m	R_e	30 Ω
m_2	0.5 kg	k_m	0.09 Wb
k_{b2}	1×10^{-6} kgm^2/s^2	k_{b1}	1×10^{-6} kgm^2/s^2
b_{b2}	1×10^{-1} kgm^{2}rad/s	b_{b1}	1×10^{-1} kgm^{2}rad/s
k_2	0.45 Vs/rad	k_1	0.0045 Vs/rad
R_2	6.96Ω	R_1	18Ω
L_2	6.031×10^{-1} H	L_1	6.031×10^{-1} H
R	l_{c2} m	V_ω	5 m/s
ρ	1.225 kg/m^3	β	0.5 rad
g	9.81 m/s^2		

electronic circuit to save the real data of the electric voltage, electric current, blades position, and tower position.

The interpolation neural network learns the behavior considering real data of the inputs and states of the wind turbine behavior, the eight inputs for the nonlinear behavior are denoted as $z_1(k) = u_{1r}$, $z_2(k) = u_{2r}$, $z_3(k) = x_{1r}$, $z_4(k) = x_{2r}$,

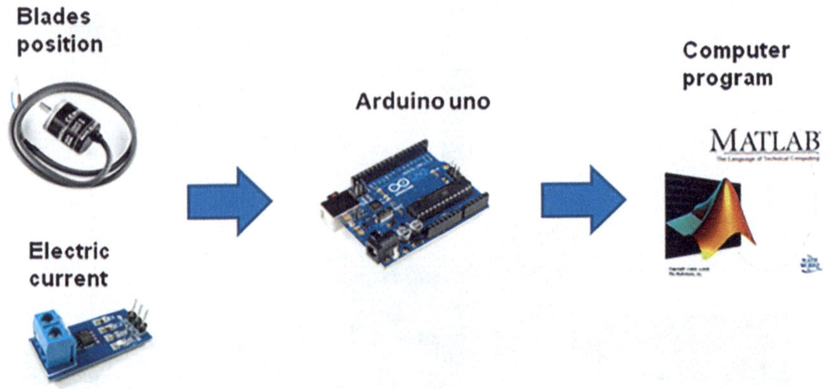

Fig. 4.5 Electronic circuit to save the real data

$z_5(k) = x_{3r}$, $z_6(k) = x_{4r}$, $z_7(k) = x_{5r}$, and $z_8(k) = x_{6r}$, and the target output is denoted as $\widehat{y}(k) = y$. The root mean square error is used for the comparison results [2, 32, 37]:

$$RMSE = \left(\frac{1}{T} \sum_{k=1}^{T} e^2(k) \right)^{\frac{1}{2}}, \tag{4.14}$$

where T is the iterations number, and $e(k) = e_{FS}(k)$ is the error of the fuzzy slopes model, or $e(k) = e_{NN}(k)$ is the error of the interpolation neural network of (4.9).

3.1 *Experiment 1*

Experiment 1 considers the first movement of the wind turbine described as follows: (1) From 0 s to 2 s, both inputs are fed; consequently, the tower moves far from the maximum air intake, the generator current is decreased, and the wind turbine blades stop moving, (2) from 2 s to 4 s, both inputs are not fed; consequently, current is not generated, and both the tower and wind turbine blades do not move, (3) from 4 s to 6 s, both inputs are fed, but the air intake is positive and tower voltage is negative; consequently, the tower returns to the maximum air intake, the generator current is increased, and the wind turbine blades move, (4) from 6 s to 8 s, both inputs are not fed; consequently, current is not generated, and the tower and wind turbine blades do not move. The described behavior is repeated three times for the learning and once for the testing; consequently, 8412 data are used for the training and 2804 data are used for the testing.

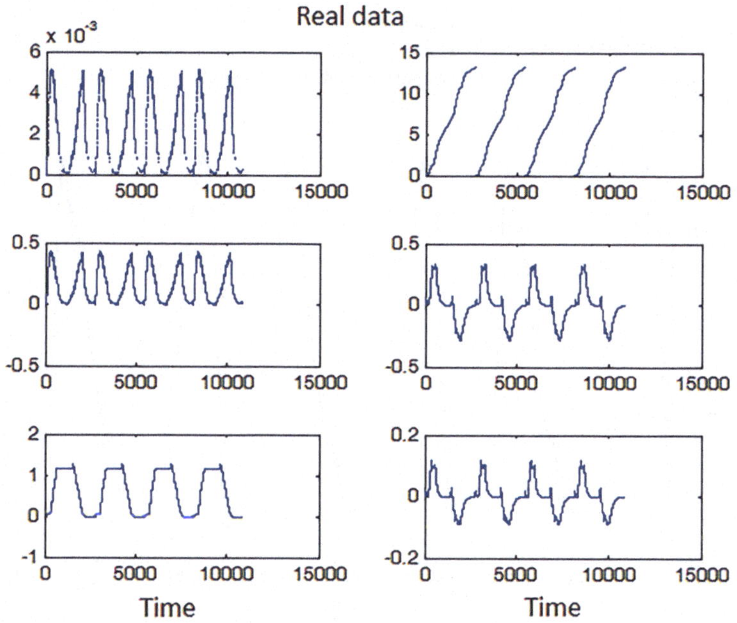

Fig. 4.6 Incomplete data for experiment 1

The fuzzy slopes model is used with parameters $n = 8$, $m = 4$, $v_i(1) = \text{rand}$, $c_{ij}(1) = \text{rand}$, $\sigma_{ij}(1) = 10\,\text{rand}$, $h = 1 \times 10^{-7}$, and rand is a random number between 0 and 1.

The interpolation neural network of (4.1)–(4.4), (4.8)–(4.10) is used with parameters $n = 8$, $m = 4$, $\alpha_0 = 0.5$, $V_j(1) = \text{rand}$, $W_{ij}(1) = \text{rand}$, $h = 1 \times 10^{-7}$, rand is a random number between 0 and 1.

Figure 4.6 shows the incomplete data for the states of the wind turbine behavior. Figure 4.7 shows the modeling of the wind turbine behavior using the fuzzy slopes model and interpolation neural network for the training. Figure 4.8 shows the modeling of the wind turbine behavior using the fuzzy slopes model and interpolation neural network for the testing. Table 4.2 shows the root mean square error for the fuzzy slopes model and interpolation neural network.

The iterations' number is shown instead of the time in seconds to guarantee that in this research incomplete data are employed. From Fig. 4.7 and Table 4.2, it is shown that the interpolation neural network is the best for the training of the wind turbine behavior because the RMSE of the above algorithm is the smallest one. The training could be used for online designs such as the control, prediction, or fault detection. From Fig. 4.8 and Table 4.2, it is shown that the interpolation neural network is the best for the testing of the wind turbine behavior because the RMSE of the above algorithm is the smallest one. The testing could be used for offline designs such as the pattern recognition or classification.

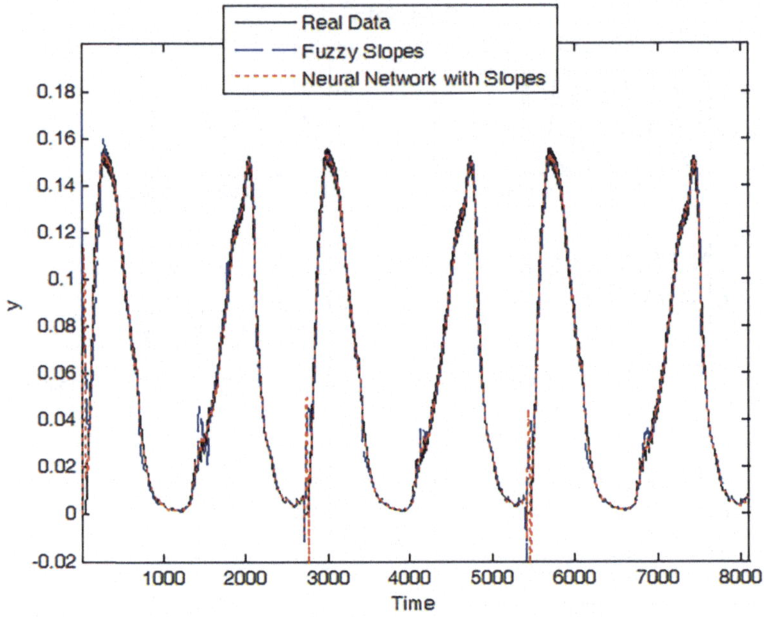

Fig. 4.7 Modeling for the training of experiment 1

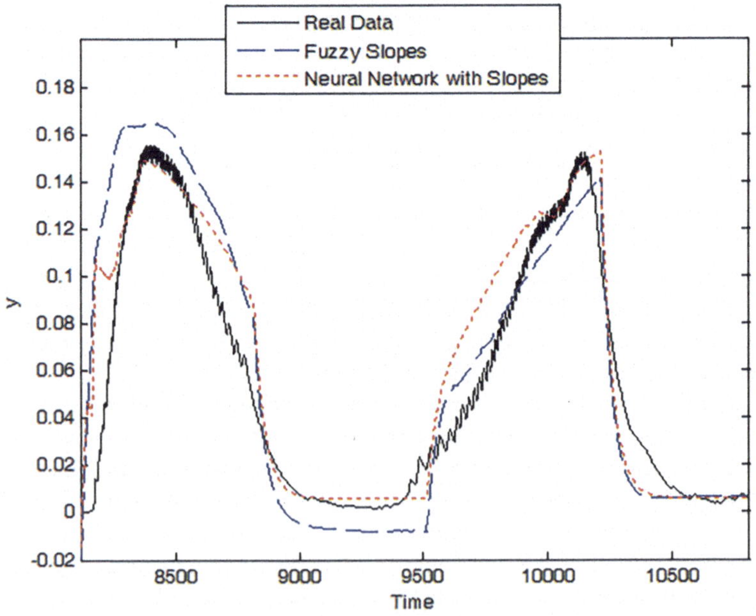

Fig. 4.8 Modeling for the testing of experiment 1

Table 4.2 Comparison of the errors for experiment 1

	RMSE for training	RMSE for testing
Fuzzy Slopes Model	0.0065	0.0086
Interpolation Neural Network	0.0049	0.0071

3.2 Experiment 2

Experiment 2 considers the second movement of the wind turbine described as follows: (1) From 0 s to 2 s, the input air is fed, and the tower input is not fed; consequently, the tower remains in the maximum air intake, the generator current is maximum, and the wind turbine blades have motion, (2) from 2 s to 4 s, the air is not fed, and the tower input is fed; consequently, current is not generated, the tower moves far from the maximum air intake, and the wind turbine blades do not have motion, (3) from 4 s to 6 s, the air is fed, and the tower input is not fed; consequently, the tower does not move, the generator current is minimum, and the wind turbine blades almost do not move, (4) from 6 s to 8 s, the air is not fed, and the tower input is fed with a negative voltage; consequently, current is not generated, the tower returns to the maximum air intake, and the wind turbine blades do not have motion. The described behavior is repeated three times for the learning and once for the testing; consequently, 8412 data are used for the training and 2804 data are used for the testing.

The fuzzy slopes model is used with parameters $n = 8$, $m = 4$, $v_i(1) = \text{rand}$, $c_{ij}(1) = \text{rand}$, $\sigma_{ij}(1) = 10\,\text{rand}$, $h = 1 \times 10^{-7}$, and rand is a random number between 0 and 1.

The interpolation neural network of (4.1)–(4.4), (4.8)–(4.10) is used with parameters $n = 8$, $m = 4$, $\alpha_0 = 0.5$, $V_j(1) = \text{rand}$, $W_{ij}(1) = \text{rand}$, $h = 5 \times 10^{-8}$, and rand is a random number between 0 and 1.

Figure 4.9 shows the incomplete data for the states of the wind turbine behavior. Figure 4.10 shows the modeling of the wind turbine behavior using the fuzzy slopes model and interpolation neural network for the training. Figure 4.11 shows the modeling of the wind turbine behavior using the fuzzy slopes model and interpolation neural network for the testing. Table 4.3 shows the root mean square error for the fuzzy slopes model and interpolation neural network.

The iterations' number is shown instead of the time in seconds to guarantee that in this research incomplete data are employed. From Fig. 4.10 and Table 4.3, it is shown that the interpolation neural network is the best for the training of the wind turbine behavior because the RMSE of the above algorithm is the smallest one. The training could be used for online designs such as the control, prediction, or fault detection. From Fig. 4.11 and Table 4.3, it is shown that the interpolation neural network is the best for the testing of the wind turbine behavior because the RMSE of the above algorithm is the smallest one. The testing could be used for offline designs such as the pattern recognition or classification.

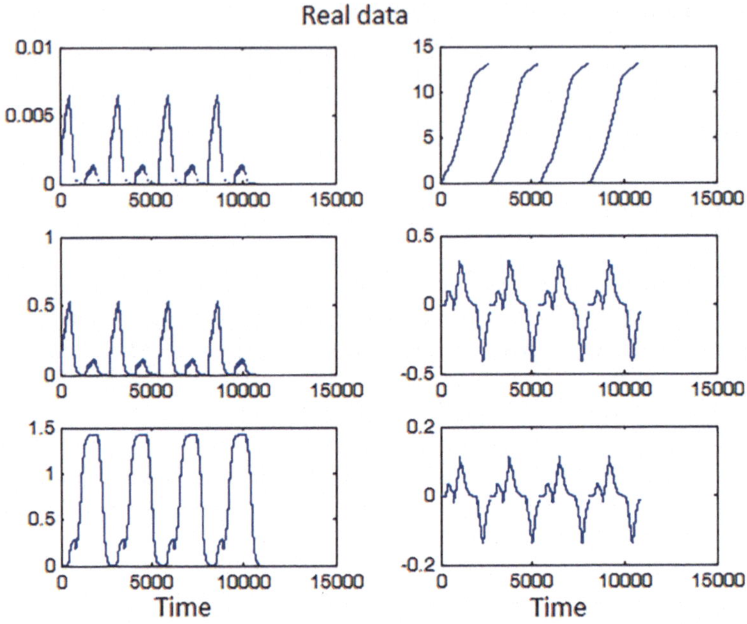

Fig. 4.9 Incomplete data for experiment 2

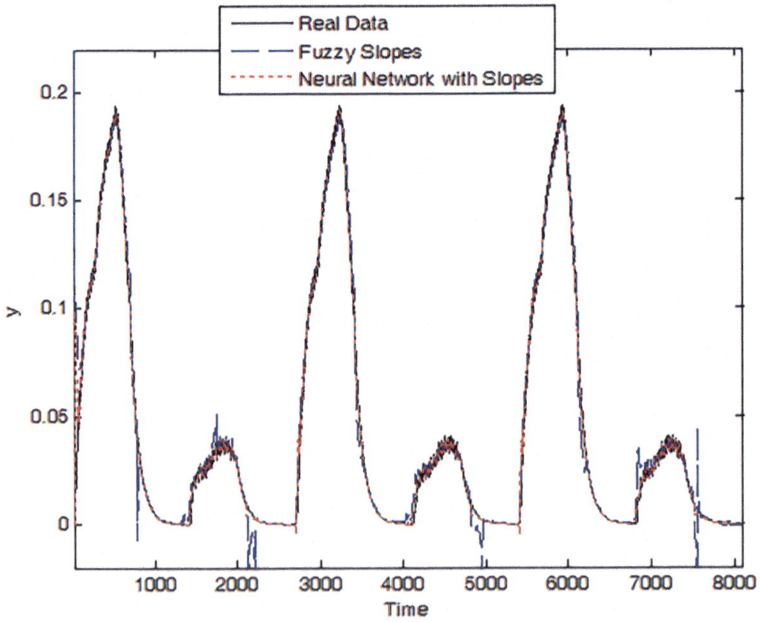

Fig. 4.10 Modeling for the training of experiment 2

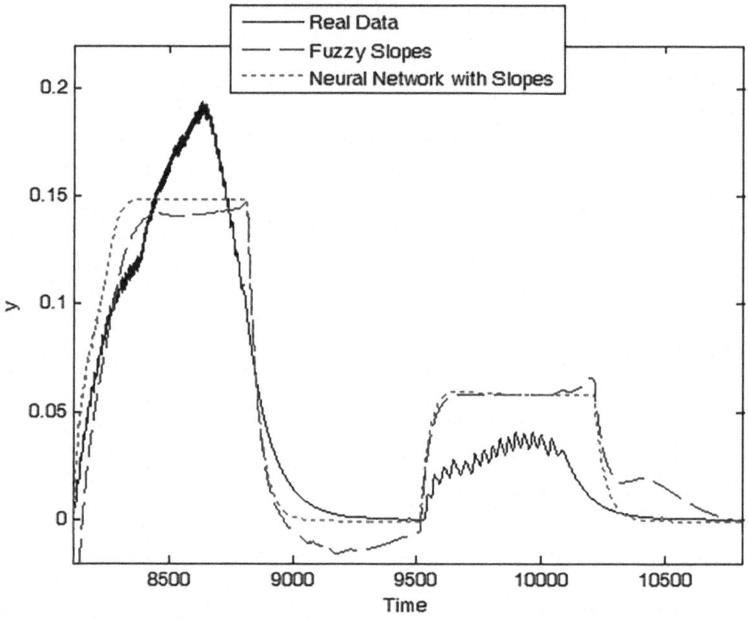

Fig. 4.11 Modeling for the testing of experiment 2

Table 4.3 Comparison of the errors for experiment 2

	RMSE for training	RMSE for testing
Fuzzy Slopes Model	0.0065	0.0086
Interpolation Neural Network	0.0035	0.0077

Remark 4.7 Choosing an appropriate number of hidden neurons is important in the behavior, because too many neurons result in a complex system that may be unnecessary for the problem, and it can cause overfitting [2], whereas too few neurons produce a less powerful system that may be insufficient to achieve the objective. The number of hidden neurons is considered as a design parameter, and it is determined based on the trial-and-error method.

4 Concluding Remarks

In this chapter, the interpolation neural network was introduced. The interpolation algorithm was applied to build an estimation of the nonlinear behaviors when only some points of the real behavior with incomplete data were available. After the interpolation algorithm obtained the estimation of the nonlinear behaviors, the neural network was employed to learn the output nonlinear behavior considering only the outputs of the interpolation model instead of the real data inputs and output. The importance of the neural network is that while the interpolation algorithm only estimates the nonlinear behaviors, the neural network learns the output behavior. The proposed interpolation neural network was compared with a fuzzy slopes model for the modeling of the wind turbine behavior, giving that the first algorithm provides higher accuracy compared to the other. The proposed technique could be used in control, prediction, pattern recognition, classification, or fault detection. As a future research, the proposed strategy will be used for the control design.

References

1. M. Pratama, M.J. Er, X. Li, R.J. Oentaryo, E. Lughofer, I. Arifin, Data driven modeling based on dynamic parsimonious fuzzy neural network. Neurocomputing **110**, 18–28 (2013)
2. J.J. Rubio, Evolving intelligent algorithms for the modelling of brain and eye signals. Appl. Soft Comput. **14**(B), 259–268 (2014)
3. J.J. Rubio, P. Angelov, J. Pacheco, An uniformly stable backpropagation algorithm to train a feedforward neural network. IEEE Trans. Neural Netw. **22**(3), 356–366 (2011)
4. H. Toubakh, M. Sayed-mouchaweh, E. Duviella, Advanced pattern recognition approach for fault diagnosis of wind turbines, in *12th International Conference on Machine Learning and Applications* (2013), pp. 368–373
5. M. Pratama, S.G. Anavatti, M.J. Er, E.D. Lughofer, pClass: An effective classifier for streaming examples. IEEE Trans. Fuzzy Syst. **23**(2), 369–386 (2015)
6. M. Pratama, J. Lu, G. Zhang, Evolving type-2 fuzzy classifier. IEEE Trans. Fuzzy Syst. **24**(3), 574–589 (2015). https://doi.org/10.1109/TFUZZ.2015.2463732
7. E. Duviella, L. Serir, M. Sayed-Mouchaweh, An evolving classification approach for fault diagnosis and prognosis of a wind farm, in *Conference on Control and Fault-Tolerant Systems (SysTol)* (2013), pp. 377–382
8. A. Lemos, W. Caminhas, F. Gomide, Adaptive fault detection and diagnosis using an evolving fuzzy classifier. Inf. Sci. **220**, 64–85 (2013)
9. E. Lughofer, E. Weigl, W. Heidl, C. Eitzinger, T. Radauer, Integrating new classes on the fly in evolving fuzzy classifier designs and its application in visual inspection. Appl. Soft Comput. **35**, 558–582 (2015)
10. A. Bouchachia, An evolving classification cascade with self-learning. Evol. Syst. **1**(3), 143–160 (2010)
11. F. Bordignon, F. Gomide, Uninorm based evolving neural networks and approximation capabilities. Neurocomputing **127**, 13–20 (2014)
12. A. Bouchachia, Learning with hybrid data, in *Proceedings of the Fifth International Conference on Hybrid Intelligent Systems* (2005), pp. 1–6
13. C. Cernuda, E. Lughofer, P. Hintenaus, W. Marzinger, T. Reischer, M. Pawliczek, J. Kasberger, Hybrid adaptive calibration methods and ensemble strategy for prediction of cloud point in melamine resin production. Chemom. Intell. Lab. Syst. **126**, 60–75 (2013)
14. N.V. Chawla, K.W. Bowyer, L.O. Hall, W.P. Kegelmayr, Smote: synthetic minority over-sampling technique. J. Artif. Intell. Res. **16**, 321–357 (2002)
15. L. Hartert, M. Sayed-Mouchaweh, Dynamic supervised classification method for online monitoring in non-stationary environments. Neurocomputing **126**, 118–131 (2014)
16. J.A. Iglesias, A. Tiemblo, A. Ledezma, A. Sanchis, Web news mining in an evolving framework. Inf. Fusion (2015). http://dx.doi.org/10.1016/j.inffus.2015.07.004
17. F.J. Ordoñez, J.A. Iglesias, P. de Toledo, A. Ledezma, A. Sanchis, Online activity recognition using evolving classifiers. Expert Syst. Appl. **40**, 1248–1255 (2013)

18. P. Kazienko, E. Lughofer, B. Trawinski, Hybrid and ensemble methods in machine learning J.UCS special issue. J. Univ. Comput. Sci. **19**(4), 457–461 (2013)
19. D. Leite, P. Costa, F. Gomide, Evolving granular neural networks from fuzzy data streams. Neural Netw. **38**, 1–16 (2013)
20. E. Lughofer, Hybrid active learning for reducing the annotation effort of operators in classification systems. Pattern Recogn. **45**, 884–896 (2012)
21. L. Maciel, F. Gomide, R. Ballini, Enhanced evolving participatory learning fuzzy modeling: an application for asset returns volatility forecasting. Evol. Syst. **5**, 75–88 (2014)
22. A. Nuñez, B. De Schutter, D. Saez, I. Skrjanc, Hybrid-fuzzy modeling and identification. Appl. Soft Comput. **17**, 67–78 (2014)
23. M. Pratama, S.G. Anavatti, P.P. Angelov, E. Lughofer, PANFIS: a novel incremental learning machine. IEEE Trans. Neural Networks Learn. Syst. **25**(1), 55–68 (2014)
24. M. Pratama, S.G. Anavatti, E. Lughofer, GENEFIS: toward an effective localist network. IEEE Trans. Fuzzy Syst. **22**(3), 547–562 (2014)
25. M. Pratama, S.G. Anavatti, J. Lu, Recurrent classifier based on an incremental meta-cognitive-based scaffolding algorithm. IEEE Trans. Fuzzy Syst. **23**(6), 2048–2066 (2015). https://doi.org/10.1109/TFUZZ.2015.2402683
26. R. Rosa, F. Gomide, R. Ballini, Evolving hybrid neural fuzzy network for system modeling and time series forecasting, in *12th International Conference on Machine Learning and Applications* (2013), pp. 1–6
27. I. Cruz-Vega, W. Yu, Multiple fuzzy neural networks modeling with sparse data. Neurocomputing **73**, 2446–2453 (2010)
28. M. Elad, Sparse and redundant representation modeling-what next? IEEE Signal Process Lett. **19**(12), 922–928 (2012)
29. L. Tao, E. Elhamifar, S. Khudanpur, G.D. Hager, R. Vidal, Sparse hidden Markov models for surgical gesture classification and skill evaluation, in *Lecture Notes in Artificial Intelligence* (1012), pp. 167–177
30. S. Zhang, Y. Zhan, M. Dewan, J. Huang, D.N. Metaxas, X.S. Zhou, Towards robust and effective shape modeling: sparse shape composition. Med. Image Anal. **16**, 265–277 (2012)
31. L.W. Zhong, J.T. Kwok, Efficient sparse modeling with automatic feature grouping. IEEE Trans. Neural Networks Learn. Syst. **23**(9), 1436–1447 (2012)
32. J.J. Rubio, D.M. Vazquez, D. Mujica-Vargas, Acquisition system and approximation of brain signals. IET Sci. Meas. Technol. **7**(4), 232–239 (2013)
33. J.J. Rubio, Interpolation neural network model of a manufactured wind turbine. Neural Comput. Appl. **28**(8), 2017–2028 (2017)
34. L.X. Wang, *A Course in Fuzzy Systems and Control*. ISBN: 0-13-540882-2 (1997)
35. J.J. Rubio, Fuzzy slopes model of nonlinear systems with sparse data. Soft Comput. **19**(12), 3507–3514 (2015). https://doi.org/10.1007/s00500-014-1289-6
36. J.J. Rubio, L.A. Soriano, W. Yu, Dynamic model of a wind turbine for the electric energy generation. Math. Probl. Eng. **2014**, 1–8 (2014)
37. A. Marques Silva, W. Caminhas, A. Lemos, F. Gomide, A fast learning algorithm for evolving neo-fuzzy neuron. Appl. Soft Comput. **14**(B), 194–209 (2014)

Chapter 5
Uniform Stable Radial Basis Function Neural Network for the Prediction in Two Mechatronic Processes

1 Introduction

Neural networks are some kind of intelligent techniques which have been employed for the prediction, pattern recognition, modeling, control, and classification in the mechatronic processes.

There is some research about the intelligent techniques. In [1], a learning approach to train uninorm-based hybrid neural networks is considered. Nature-inspired algorithms are described in [2]. In [3], the utilization of nature-inspired algorithms in sports is detailed. Intelligent algorithm to save human lives is addressed in [4]. In [5], and a granular neural network framework is introduced. The time varying coefficients in a model are approximated in [6]. In [7], a clustering method is introduced. In [8], a hybrid active learning strategy is proposed. The overlapping of radial basis functions inside a cerebellar model arithmetic computer is studied in [9]. In [10], the game theoretical models are studied. A hybrid dynamic classifier is addressed in [11]. In [12], each part of a pump system is modeled. A radial basis function neural network of motion control is discussed in [13]. In [14], radial basis function neural networks to perform interval forecasting of the future wind speed are proposed.

From the above proposals, [6, 9, 12, 13], and [14] consider the radial basis function neural networks, and it shows that this network is novel and actual research. Therefore, new studies in this kind of neural network should be of great interest.

The backpropagation with variable learning steps, mentioned in [15–18], and [19], is employed for the learning of a feedforward neural network. It is an efficient algorithm; therefore, it would be good to modify this approach to be employed in a radial basis function neural network.

In this chapter, the algorithm used for the learning of the one hidden layer neural network is modified to be applied in a radial basis function neural network. The problem is difficult because the one hidden layer neural network uses sigmoid functions, while the radial basis function neural network uses Gaussian functions.

J. de J. Rubio, *Stability Analysis of Neural Networks and Evolving Intelligent Systems*, https://doi.org/10.1007/978-3-031-87282-2_5

The Gaussian function can be adapted better to the changing behavior of the systems than the sigmoid function for two reasons: (1) The first has one parameter for the width and two for the centers, while the second only uses the two parameters for the centers, and (2) the first can learn positive and negative values, while the second only can learn positive values. Therefore, the design of the addressed algorithm is more complex, but it is more effective for the learning.

On the other hand, there is some research about the stable intelligent systems. In [20], a new filter with a finite impulse response structure of models. In [21], a passive and exponential filter of switched Hopfield neural networks is used. The non-divergence of the original discrete-time algorithms is analyzed in [15]. In [22], [23], and [24], stable neuro controllers of nonlinear systems are designed. The stability of a Markov jump recurrent neural network and a hierarchical hybrid neural network are analyzed in [25] and [26]. In [27], global exponential stability of the complex-valued recurrent neural networks is investigated. Global stability of the complex-valued neural networks with discrete-time delay is studied in [28]. In [29], a class of inertial neural networks with delays is considered. A stable complex delayed dynamic network is developed in [30]. In [31], a stable directed complex dynamic network is suggested. The concept of impulsive time window is proposed in [32].

From the aforementioned works, in [20, 21], the stability of continuous-time neural networks is studied, in [15, 22, 23], the stability of the backpropagation algorithm is introduced, in [25, 26], the stability of continuous-time neural networks is described, and in [27–29], the stability of neural networks with delays is analyzed. The aforementioned research shows that the stability analysis of algorithms for neural networks is an actual issue. In this chapter, the uniform stability of the before mentioned method is assured.

In this chapter the algorithm used for the learning of the one hidden layer neural network is modified to be applied in a radial basis function neural network where its stability is assured. To reach the stability and convergence of the error to a small value, a time varying learning parameter is introduced. It assures an acceptable behavior of the algorithm to some undesired situations such as the disturbances or faults.

The rest of this chapter is organized as follows. In Sect. 2, the radial basis function neural network is introduced. In Sect. 3, the radial basis function neural network is linearized. In Sect. 4, the addressed algorithm for the learning of a radial basis function neural network is designed. In Sect. 5, the stability, convergence, and boundedness of parameters for the aforementioned strategy are assured. In Sect. 6, the focused method is summarized. In Sect. 7, the mentioned algorithm is compared with the uniform stable neural network for the two processes. Section 8 describes the conclusions and future research alternatives.

2 Radial Basis Function Neural Network

Consider the following unknown discrete-time multiple input multiple output mechatronic process:

$$y_l(k) = f_l\left[X_k\right], \tag{5.1}$$

where $i = 1, \ldots, N$, $l = 1, \ldots, O$, $X_k = [x_1(k) \ldots, x_i(k), \ldots, x_N(k)]^T \in \Re^{N \times 1}$ is the input vector, N is the input number, O is the output number, $x_i(k) \in \Re$ and $y_l(k) \in \Re$ are the inputs and outputs of the plant, and f is an unknown and smooth nonlinear function, $f_l \in C^{\infty}$.

The outputs of the radial basis function neural network with one hidden layer are as follows:

$$\begin{aligned}
\widehat{y}_l(k) &= \frac{d_l(k)}{g(k)} = \frac{\sum\limits_{j=1}^{M} \widehat{r}_{jl}(k)\alpha_j(u_j(k))}{\sum\limits_{j=1}^{M} \alpha_j(u_j(k))}, \\
\alpha_j(u_j(k)) &= e^{-u_j^2(k)}, \\
u_j(k) &= \sum_{i=1}^{N} \widehat{s}_{ij}(k)\left[x_i(k) - \widehat{t}_i(k)\right],
\end{aligned} \tag{5.2}$$

where $i = 1, \ldots, N$, $j = 1, \ldots, M$, $l = 1, \ldots, O$, $x_i(k) \in \Re$ and $\widehat{y}_l(k) \in \Re$ are the inputs and outputs of the neural network, $\widehat{r}_{jl}(k) \in \Re$, $\widehat{s}_{ij}(k) \in \Re$, $\widehat{t}_i(k) \in \Re$ are the weights of the output and hidden layers and centers of the neural network, respectively, $\alpha_j(u_j(k)) \in \Re$ is a nonlinear function, $u_j(k) \in \Re$ is the addition function, M is the neuron number in the hidden layer, and O is the output number. Figure 5.1 shows the architecture of the radial basis function neural network.

Remark 5.1 Mechatronic processes of the form (5.1) are general because they are known as multiple input multiple output processes. Other general processes are the delayed processes or the multiple inputs multiple states where the design methods are similar.

3 Linearization of the Radial Basis Function Neural Network

The linearization of the radial basis function neural network is required for the algorithm design and for the stability analysis.

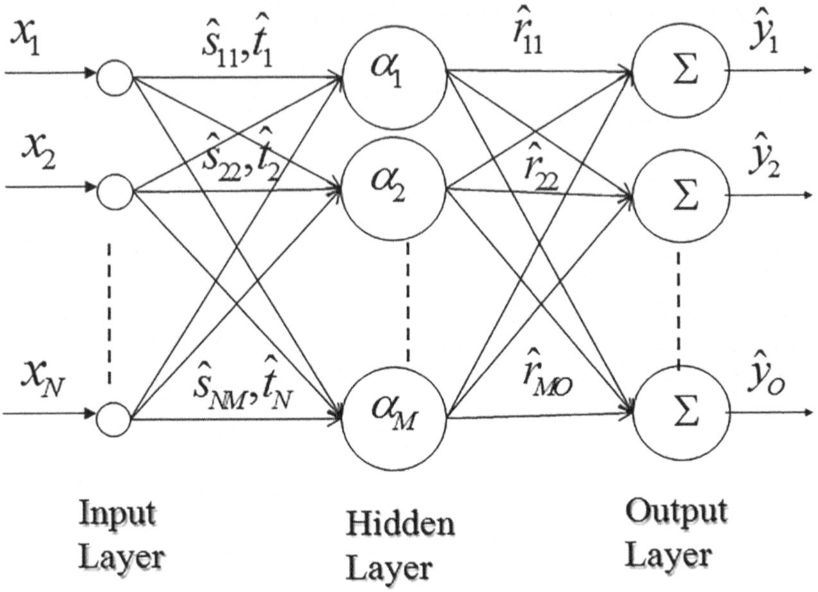

Fig. 5.1 Architecture of the neural network

According to the Stone-Weierstrass theorem, the unknown nonlinear function f of (5.1) is approximated as

$$
\begin{aligned}
y_l(k) &= \frac{d_{l*}}{g_*} + \in_{lf} = \frac{\sum_{j=1}^{M} r_{jl*}\alpha_{j*}}{\sum_{j=1}^{M} \alpha_{j*}} + \in_{lf}, \\
\alpha_{j*} &= e^{-u_{j*}^2}, \\
u_{j*} &= \sum_{i=1}^{N} s_{ij*}\left[x_i(k) - t_{i*}\right],
\end{aligned}
\tag{5.3}
$$

where $\in_{lf} = y_l(k) - \frac{d_{l*}}{g_*} \in \Re$ are the modeling errors, $\alpha_{j*} \in \Re$, $r_{jl*} \in \Re$, $s_{ij*} \in \Re$, and $t_{i*} \in \Re$ are the optimal parameters that can minimize the modeling errors $\in_{lf}$. In the case of three independent variables, a function has a Taylor series as follows:

$$
\begin{aligned}
f_l(\omega_1, \omega_2, \omega_3) &= f_l(\omega_{1^0}, \omega_{2^0}, \omega_{3^0}) + \left(\omega_1 - \omega_{1^0}\right) \frac{\partial f_l(\omega_1, \omega_2, \omega_3)}{\partial \omega_1} \\
&+ \left(\omega_2 - \omega_{2^0}\right) \frac{\partial f_l(\omega_1, \omega_2, \omega_3)}{\partial \omega_2} + \left(\omega_3 - \omega_{3^0}\right) \frac{\partial f_l(\omega_1, \omega_2, \omega_3)}{\partial \omega_3} + \varsigma_{lf},
\end{aligned}
\tag{5.4}
$$

where $\varsigma_{lf} \in \Re$ is the remainder of the Taylor series. ω_1, ω_2, and ω_3 correspond to $\widehat{r}_{jl}(k) \in \Re$, $\widehat{s}_{ij}(k) \in \Re$, and $\widehat{t}_i(k) \in \Re$, ω_{1^0}, ω_{2^0}, and ω_{3^0} correspond to $r_{jl*} \in \Re$, $s_{ij*} \in \Re$, and $t_{i*} \in \Re$, define $\widetilde{r}_{jl}(k) = \widehat{r}_{jl}(k) - r_{jl*} \in \Re$, $\widetilde{s}_{ij}(k) = \widehat{s}_{ij}(k) - s_{ij*} \in \Re$, and $\widetilde{t}_i(k) = \widehat{t}_i(k) - t_{i*} \in \Re$; consequently, the Taylor series is applied to linearize (5.2) as follows:

$$\begin{aligned}\frac{d_l(k)}{g(k)} = \frac{d_{l*}}{g_*} &+ \sum_{j=1}^{M} \widetilde{r}_{jl}(k)\frac{\partial \frac{d_l(k)}{g(k)}}{\partial \widehat{r}_{jl}(k)} + \sum_{i=1}^{N}\sum_{j=1}^{M} \widetilde{s}_{ij}(k)\frac{\partial \frac{d_l(k)}{g(k)}}{\partial \widehat{s}_{ij}(k)} \\ &+ \sum_{i=1}^{N} \widetilde{t}_i(k)\frac{\partial \frac{d_l(k)}{g(k)}}{\partial \widehat{t}_i(k)} + \varsigma_{lf},\end{aligned} \tag{5.5}$$

where $\frac{\partial \frac{d_l(k)}{g(k)}}{\partial \widehat{r}_{jl}(k)} \in \Re$, $\frac{\partial \frac{d_l(k)}{g(k)}}{\partial \widehat{s}_{ij}(k)} \in \Re$, and $\frac{\partial \frac{d_l(k)}{g(k)}}{\partial \widehat{t}_i(k)} \in \Re$; please note that $d_l(k) = \sum_{j=1}^{M} \widehat{r}_{jl}(k)\alpha_j(u_j(k)) \in \Re$, $g(k) = \sum_{j=1}^{M} \alpha_j(u_j(k)) \in \Re$. As all the parameters are scalars, the Taylor series can be utilized. Considering (5.2) and using the chain rule, it gives

$$\frac{\partial \frac{d_l(k)}{g(k)}}{\partial \widehat{r}_{jl}(k)} = R_j(k) = \frac{\alpha_j(u_j(k))}{g(k)}, \tag{5.6}$$

where $\alpha_j(u_j(k)) = e^{-u_j^2(k)}$ and $u_j(k)$ are given in (5.2). Subsequently,

$$\begin{aligned}\frac{\partial \frac{d_l(k)}{g(k)}}{\partial \widehat{s}_{ij}(k)} &= S_{ijl}(k) \\ &= \frac{[\gamma_j(u_j(k))\widehat{r}_{jl}(k) + \gamma_j(u_j(k))\widehat{y}_l(k)][\widehat{t}_i(k) - x_i(k)]}{g(k)},\end{aligned} \tag{5.7}$$

where $\gamma_j(u_j(k)) = 2u_j(k)\alpha_j(u_j(k)) \in \Re$. And

$$\begin{aligned}\frac{\partial \frac{d_l(k)}{g(k)}}{\partial \widehat{t}_i(k)} &= T_{il}(k) \\ &= \frac{\widehat{s}_{ij}(k)[\gamma_j(u_j(k))\widehat{r}_{jl}(k) + \gamma_j(u_j(k))\widehat{y}_l(k)]}{g(k)}.\end{aligned} \tag{5.8}$$

Substituting $\frac{\partial \frac{d_l(k)}{g(k)}}{\partial \widehat{r}_{jl}(k)}$ of (5.6), $\frac{\partial \frac{d_l(k)}{g(k)}}{\partial \widehat{s}_{ij}(k)}$ of (5.7), and $\frac{\partial \frac{d_l(k)}{g(k)}}{\partial \widehat{t}_i(k)}$ of (5.8) into (5.5), it gives

$$\begin{aligned}\frac{d_l(k)}{g(k)} = \frac{d_{l*}}{g_*} &+ \sum_{j=1}^{M} \widetilde{r}_{jl}(k)R_j(k) + \sum_{i=1}^{N}\sum_{j=1}^{M} \widetilde{s}_{ij}(k)S_{ijl}(k) \\ &+ \sum_{i=1}^{N} \widetilde{t}_i(k)T_{il}(k) + \varsigma_{lf}.\end{aligned} \tag{5.9}$$

Define the output errors $e_l(k) \in \Re$ as follows:

$$e_l(k) = \widehat{y}_l(k) - y_l(k), \tag{5.10}$$

where $y_l(k)$ and $\widehat{y}_l(k)$ are described in (5.1) and (5.2), respectively. Substituting (5.2), (5.3), and (5.10) into (5.9) gives

$$\begin{aligned} e_l(k) = &\sum_{j=1}^{M} \widetilde{r}_{jl}(k) R_j(k) + \sum_{i=1}^{N}\sum_{j=1}^{M} \widetilde{s}_{ij}(k) S_{ijl}(k) \\ &+ \sum_{i=1}^{N} \widetilde{t}_i(k) T_{il}(k) + \mu_l(k), \end{aligned} \tag{5.11}$$

where $\mu_l(k) = \varsigma_{lf} - \epsilon_{lf}$.

4 Design of the Addressed Algorithm

In this section, the addressed algorithm is designed for the learning of a radial basis function neural network.

Theorem 5.1 *The addressed algorithm that is the updating function of the radial basis function neural network (5.2) for the learning of the mechatronic process (5.1) is given as follows:*

$$\begin{aligned} \widehat{r}_{jl}(k+1) &= \widehat{r}_{jl}(k) - \eta(k) R_j(k) e_l(k), \\ \widehat{s}_{ij}(k+1) &= \widehat{s}_{ij}(k) - \eta(k) S_{ijl}(k) e_l(k), \\ \widehat{t}_i(k+1) &= \widehat{t}_i(k) - \eta(k) T_{il}(k) e_l(k), \end{aligned} \tag{5.12}$$

where $R_j(k)$, $S_{ijl}(k)$, and $T_{il}(k)$ are given in (5.6), (5.7), and (5.8), respectively, and $e_l(k)$ are the output errors of (5.10).

Proof *See [33] for the proof.*

Remark 5.2 The difference between the addressed algorithm and the well-known backpropagation algorithm is that the first has a time varying learning speed, while the second has constant learning speed. The difference between the focused algorithm and the time varying learning speed approach is that the second is commonly employed in a multilayer neural network, while the first is applied in the radial basis function neural network. The radial basis function neural network is more complex in the design than the multilayer neural network because the first has more parameters than the second.

Remark 5.3 The conservatism issue of the radial basis function of this study is mentioned in two parts as follows: (1) Since the radial basis function has more parameters than the multilayer, it has more computational cost because the first requires more operations, and this difference is less strong than before because now the computers make the operations very fast, and (2) some authors in the past mentioned that the radial basis function required more neurons in the hidden layer than the multilayer, but in this research, both neural networks use the same number of neurons in the hidden layer with satisfactory results.

5 Stabilization of the Addressed Algorithm

The addressed algorithm is given in (5.12) with a time varying learning speed as follows:

$$\begin{aligned} \widehat{r}_{jl}(k+1) &= \widehat{r}_{jl}(k) - \eta(k) R_j(k) e_l(k), \\ \widehat{s}_{ij}(k+1) &= \widehat{s}_{ij}(k) - \eta(k) S_{ijl}(k) e_l(k), \\ \widehat{t}_i(k+1) &= \widehat{t}_i(k) - \eta(k) T_{il}(k) e_l(k), \end{aligned} \tag{5.13}$$

where the new time varying learning speed $\eta(k)$ is

$$\eta(k) = \frac{\eta_0}{2\left(\frac{1}{2} + \sum_{j=1}^{M} R_j^2(k) + \sum_{i=1}^{N}\sum_{j=1}^{M} S_{ijl}^2(k) + \sum_{i=1}^{N} T_{il}^2(k)\right)},$$

where $i = 1, \ldots, N$, $j = 1, \ldots, M$, $l = 1, \ldots, O$, $R_j(k) \in \Re$ are described in (5.6), $S_{ijl}(k) \in \Re$ are described in (5.7), $T_{il}(k) \in \Re$ are described in (5.8), $e_l(k)$ are described in (5.10), $0 < \eta_0 \leq 1 \in \Re$, consequently $0 < \eta(k) \in \Re$, and it is assumed that the uncertainty is bounded, where $\overline{\mu}_l$ is the upper bound of the uncertainty $\mu_l(k)$, $|\mu_l(k)| < \overline{\mu}_l$.

Remark 5.4 $\eta(k)$ is chosen by the user as an average and bounded function such that the stability of the algorithm (5.13) can be assured. This kind of function was considered in [19], with the difference that the algorithm of this research considers three different parameters in the denominator and the previous one only considered two different parameters in the denominator.

The following theorem gives the stability of the addressed algorithm.

Theorem 5.2 *The algorithm (5.2), (5.10), and (5.13) applied for the identification of the mechatronic process (5.1) is uniformly stable, and the upper bound of the average output errors $e_{lp}^2(k)$ satisfies*

$$\limsup_{T\to\infty} \frac{1}{T} \sum_{k=2}^{T} e_{lp}^2(k) \leq \alpha_0 \overline{\mu}_l^2, \tag{5.14}$$

where $e_{lp}^2(k) = \frac{\eta(k-1)}{2} e_l^2(k-1)$, $0 < \eta_0 \leq 1 \in \Re$ *and* $0 < \eta(k) \in \Re$ *are described in (5.13),* $e_l(k)$ *are described in (5.11),* $\overline{\mu}_l$ *are the upper bound of the uncertainties* $\mu_l(k)$, $|\mu_l(k)| < \overline{\mu}_l$.

Proof See [33] for the proof.

Remark 5.5 There are two requirements to apply this algorithm for the learning of the behavior in mechatronic processes: the first is that the uncertainty $\mu_l(k)$ must be bounded, and the second is that the mechatronic process must have the structure explained by (5.1).

Remark 5.6 The bound of $\mu_l(k)$ denoted as $\overline{\mu}_l$ is not employed in the introduced algorithm (5.2), (5.10), (5.13) because it is only employed to assure its stability.

The following theorem proves that the weights of the suggested algorithm are bounded.

Theorem 5.3 *When the average error* $e_{lp}^2(k+1)$ *is bigger than the uncertainty* $\eta_0 \overline{\mu}_l^2$*, the weights' errors are bounded by the initial weights' errors as follows:*

$$\begin{aligned} & e_{lp}^2(k+1) \geq \eta_0 \overline{\mu}^2 \\ \Longrightarrow & \sum_{j=1}^{M} \widetilde{r}_{jl}^2(k) + \sum_{j=1}^{M}\sum_{i=1}^{N} \widetilde{s}_{ij}^2(k) + \sum_{i=1}^{N} \widetilde{t}_i^2(k) \\ \leq & \sum_{j=1}^{M} \widetilde{r}_{jl}^2(1) + \sum_{j=1}^{M}\sum_{i=1}^{N} \widetilde{s}_{ij}^2(1) + \sum_{i=1}^{N} \widetilde{t}_i^2(1), \end{aligned} \tag{5.15}$$

where $i = 1, \ldots, N$, $j = 1, \ldots, M$, $l = 1, \ldots, O$, $\widetilde{r}_{jl}(k)$, $\widetilde{s}_{ij}(k)$, *and* $\widetilde{t}_i(k)$ *are described in (5.4),* $\widetilde{r}_{jl}(1)$, $\widetilde{s}_{ij}(1)$, *and* $\widetilde{t}_i(1)$ *are the initial weights' errors,* $e_{lp}^2(k+1) = \frac{1}{2}\eta(k)e_l^2(k)$, $\widehat{r}_{jl}(k+1)$, $\widehat{s}_{ij}(k+1)$, $\widehat{t}_i(k+1)$, $0 < \eta_0 \leq 1 \in \Re$, *and* $0 < \eta(k) \in \Re$ *are described in (5.13),* $e_l(k)$ *are described in (5.10),* $\overline{\mu}_l$ *are the upper bound of the uncertainties* $\mu_l(k)$, $|\mu_l(k)| < \overline{\mu}_l$.

Proof *See [33] for the proof.*

Remark 5.7 From Theorem 5.2 the average output error $e_{lp}^2(k+1)$ of the introduced approach is bounded, and from Theorem 5.3 the weights' errors $\widetilde{r}_{jl}^2(k)$, $\widetilde{s}_{ij}^2(k)$, and $\widetilde{t}_i^2(k)$ are bounded, i.e., the suggested method to train a radial basis function neural network is uniformly stable in the presence of unmodeled dynamics, and the overfitting is avoided. And the output errors converge to a small zone bounded by the uncertainty $\overline{\mu}_l$.

6 The Addressed Algorithm

The addressed algorithm is as follows:

(1) Obtain the outputs of the mechatronic process $y_l(k)$ with Eq. (5.1). Note that the mechatronic process may have the structure represented by Eq. (5.1); the parameters N and O are selected according to the input and output number of this mechatronic process.
(2) Select the following parameters: $\widehat{r}_{jl}(1)$, $\widehat{s}_{ij}(1)$, and $\widehat{t}_i(1)$ as random numbers between 0 and 1, M as an integer number, and η_0 as a positive value smaller than or equal to 1; obtain the outputs of the radial basis function neural network $\widehat{y}_l(1)$ with Eq. (5.2).
(3) For each iteration k, obtain the outputs of the radial basis function neural network $\widehat{y}_l(k)$ with Eq. (5.2), also obtain the output errors $e_l(k)$ with Eq. (5.10), and update the parameters $\widehat{r}_{jl}(k+1)$, $\widehat{s}_{ij}(k+1)$, and $\widehat{t}_i(k+1)$ with Eq. (5.13).
(4) Note that the behavior of the algorithm could be improved by changing the values of η_0 or M.

Remark 5.8 The radial basis function neural network of this research has one hidden layer. A radial basis function neural network with one hidden layer is enough to approximate any nonlinear system.

7 Simulation Results

In this section, two examples are considered. In the examples, the addressed algorithm denoted as USRBFNN is applied for the prediction of the warehouse process and for the prediction of the brain behavior. In all cases, the focused method is compared with the uniform stable neural network given by Rubio et al. [19] denoted as USNN. The root mean square error (RMSE) is used for the comparison of algorithms , and it is given as follows:

$$\text{RMSE} = \left(\frac{1}{T} \sum_{k=1}^{T} \sum_{l=1}^{O} e_l^2(k) \right)^{\frac{1}{2}}, \tag{5.16}$$

where $e_l(k)$ are the output errors of (5.10), T is the iterations number, and O is the outputs number.

7.1 Example 1

In this example, the introduced algorithm is applied for the prediction of the distribution of loads that the warehouse receives from a vehicle and places in the deposits each hour. There are three kinds of objects received by the warehouse; these three kinds of objects are denoted as X, Y, and Z. The three kinds of objects are received in the warehouse each hour; the number of objects of kind X received each hour change from 0 to 5, the number of objects of kind Y received each hour change from 0 to 5, and the number of objects of kind Z received each hour change from 0 to 10. The data from 1800 iterations are used for the training, and the data for at least 200 iterations are used for the testing. The prediction is obtained for 200 iterations beforehand. One radial basis function neural network is used for the training, and the same network is used for the testing. $x_1(k) = Y(k)$ and $x_2(k) = Z(k)$ are the inputs, and $y_1(k) = X(k + 200)$ is the output for the learning of the first process. $x_3(k) = X(k)$ and $x_4(k) = Z(k)$ are the inputs, and $y_2(k) = Y(k+200)$ is the output for the learning of the second process. Finally, $x_5(k) = X(k)$ and $x_6(k) = Y(k)$ are the inputs, and $y_3(k) = Z(k+200)$ is the output for the learning of the third process.

The USRBFNN is given as (5.2), (5.10), and (5.13) with parameters $N = 6$, $O = 3$, $M = 10$, $\eta_0 = 1$, and $\widehat{r}_{jl}(1)$, $\widehat{s}_{ij}(1)$, and $\widehat{t}_i(1)$ are random numbers between 0 and 1. The USNN is given by Rubio et al. [19] with parameters $N = 6$, $O = 3$, $M = 10$, $\alpha_0 = 1$, and V_{j1} and W_{ij1} are random numbers between 0 and 1.

The comparison results for the average output errors are shown in Fig. 5.2 where in USRBFNN the final average error is 0.0035 and in USNN of [19] the final average error is 0.0088. Figure 5.3 shows the training results and Fig. 5.4 shows the testing results. Table 5.1 shows the training and testing RMSE results using (5.16).

From Figs. 5.2, 5.3, and 5.4, it is shown that the USRBFNN is better than the USNN because the signal of the first follows better the signal of the plant than the signal of the second. From Table 5.1, it can be shown that the USRBFNN obtained better accuracy when compared with the USNN because the RMSE is smaller for the first. Thus, the USRBFNN is preferable for the warehouse process.

7.2 Example 2

Here a real dataset of brain signals consisting of 1750 pairs $(x(k), y(k))$ of 35 s are used for the training and 250 pairs $(x(k), y(k))$ for 5 s are used for the testing. The alpha signal is obtained in this study because it has more probabilities to be found. The acquisition system is applied with a 28-year-old healthy man when his eyes are closed. There are three different signals received by the brain signals; these three kinds of signals are denoted as X, Y, and Z. The prediction is obtained for 250 iterations in advance. One radial basis function neural network is used for the training, and the same network is used for the testing. $x_1(k) = X(k)$, $x_2(k) = Y(k)$,

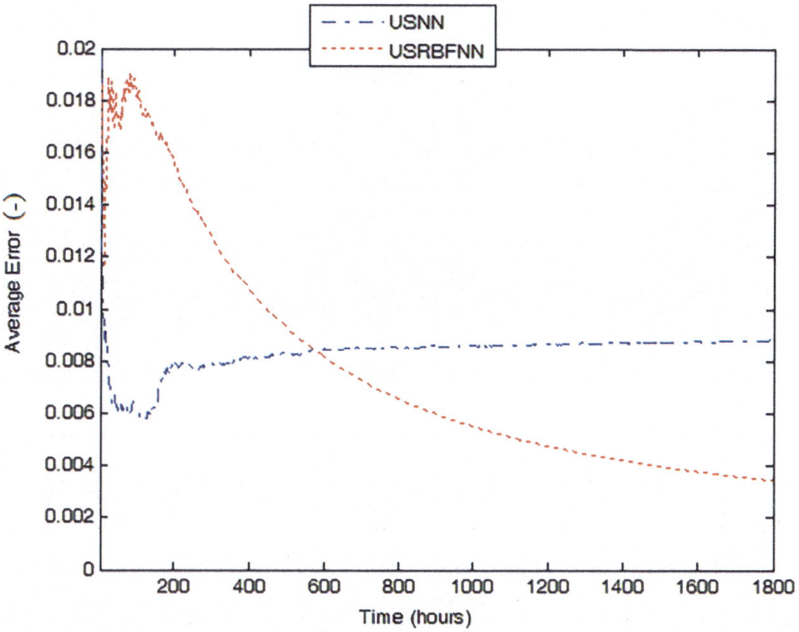

Fig. 5.2 Average learning errors for Example 1

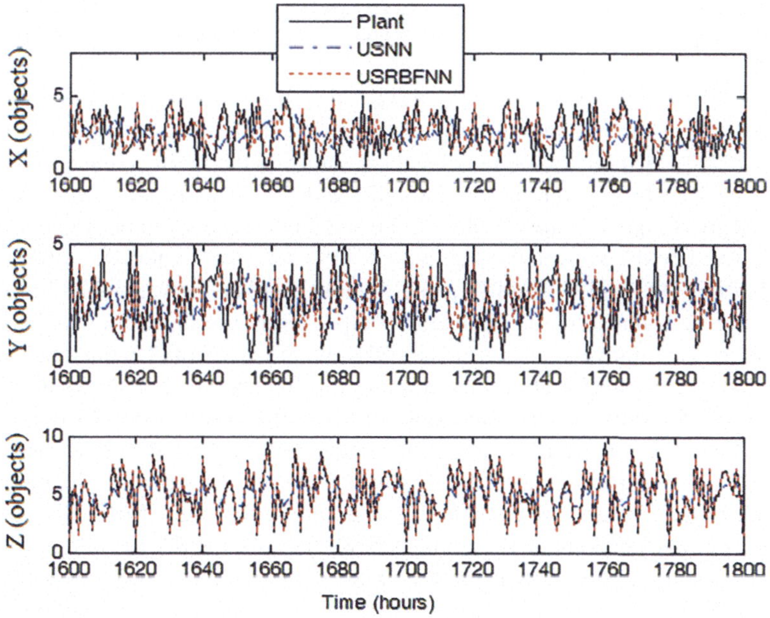

Fig. 5.3 Training results for Example 1

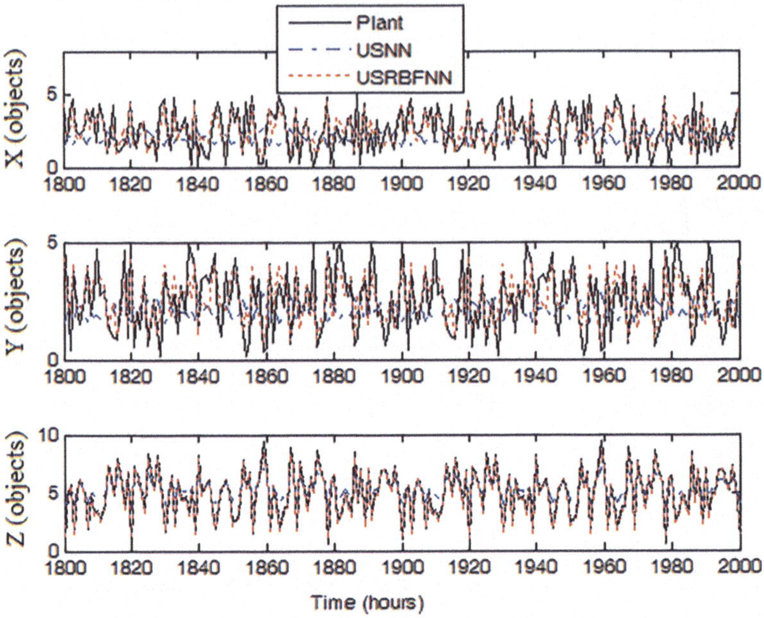

Fig. 5.4 Testing results for Example 1

Table 5.1 Results for Example 1

Strategies	Training RMSE	Testing RMSE
USRBFNN	0.1238	0.0255
USNN	0.2005	0.0604

and $x_3(k) = Z(k)$ are the inputs and $y_1(k) = X(k+250)$, $y_2(k) = Y(k+250)$, and $y_3(k) = Z(k+250)$ are the outputs for the training of the brain signals process.

The USRBFNN is given as (5.2), (5.10), and (5.13) with parameters $N = 6$, $O = 3$, $M = 10$, $\eta_0 = 1$, and $\widehat{r}_{jl}(1)$, $\widehat{s}_{ij}(1)$, and $\widehat{t}_i(1)$ are random numbers between 0 and 1. The USNN is given by Rubio et al. [19] with parameters $N = 6$, $O = 3$, $M = 10$, $\alpha_0 = 0.5$, and V_{j1} and W_{ij1} are random numbers between 0 and 1.

The comparison results for the average output errors are shown in Fig. 5.5 where in USRBFNN the final average error is 0.0016 and in USNN of [19] the final average error is 0.0370. Figure 5.6 shows the training results, and Fig. 5.7 shows the testing results. Table 5.2 shows the training and testing RMSE results using (5.16).

From Figs. 5.5, 5.6, and 5.7, it can be shown that the USRBFNN is better than the USNN because the signal of the first follows better the signal of the plant than the signal of the second. From Table 5.2, it is shown that the USRBFNN obtained better accuracy when compared with the USNN because the RMSEs are smaller for the first. Thus, the USRBFNN is preferable for the crude oil blending process.

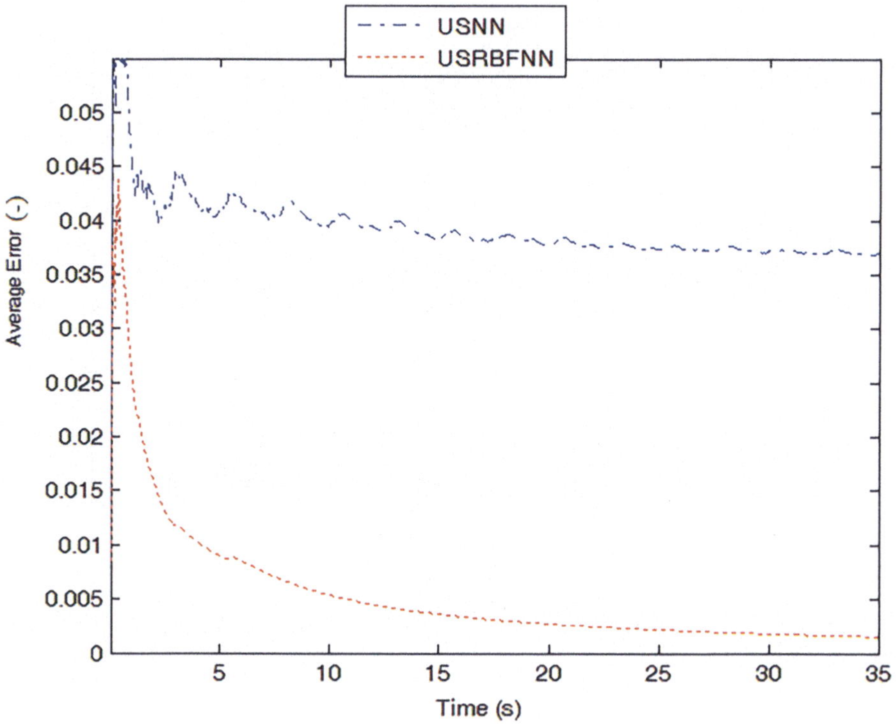

Fig. 5.5 Average learning errors for Example 2

8 Concluding Remarks

In this chapter, a novel algorithm is designed for the learning of a radial basis function neural network, and the stability, convergence, and boundedness of parameters for the addressed algorithm are assured. From the results, it was shown that the focused strategy achieves better accuracy when compared with the uniform stable neural network for the prediction of two mechatronic processes. The studied method could be used to train a neural network as was applied in this chapter, or it could be used as the parameters updating of an evolving intelligent system. As a future research, the mentioned method will be used for the control design or for the learning of evolving intelligent systems, or the properties of other interesting algorithms will be analyzed.

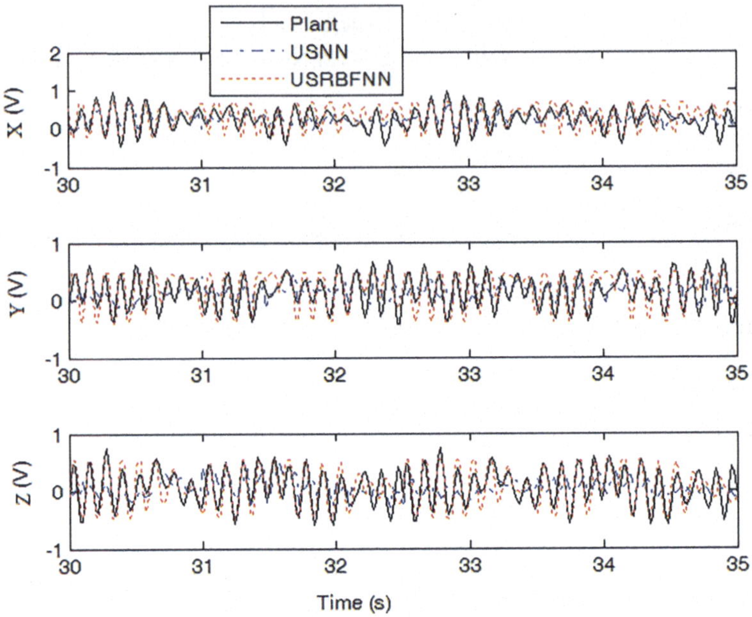

Fig. 5.6 Training results for Example 2

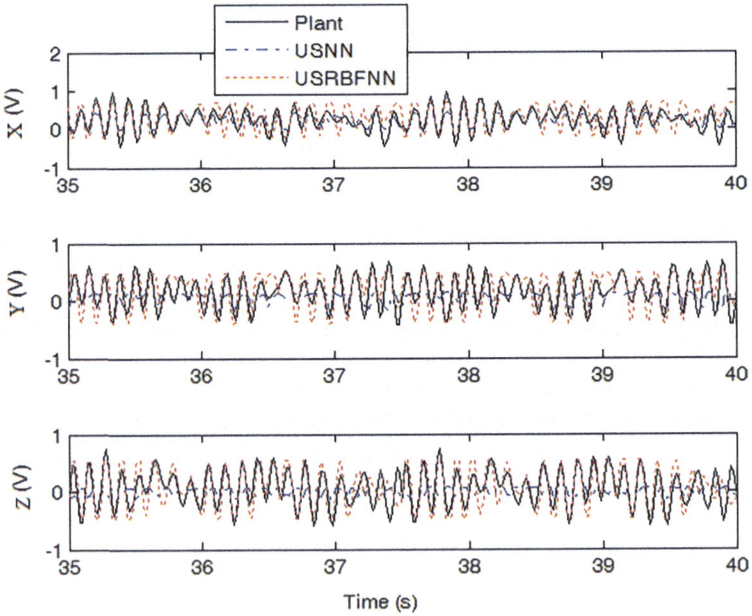

Fig. 5.7 Testing results for Example 2

Table 5.2 Results for Example 2

Strategies	Training RMSE	Testing RMSE
USRBFNN	0.2944	0.1087
USNN	0.3496	0.1195

References

1. F. Bordignon, F. Gomide, Uninorm based evolving neural networks and approximation capabilities. Neurocomputing **127**, 13–20 (2014)
2. I. Fister, M. Perc, S.M. Kamal, I. Fister, A review of chaos-based firefly algorithms: perspectives and research challenges. Appl. Math. Comput. **252**, 155–165 (2015)
3. I. Fister, K. Ljubic, P. Nagaratnam, M. Perc, I. Fister, Computational intelligence in sports: challenges and opportunities within a new research domain. Appl. Math. Comput. **262**, 178–186 (2015)
4. D. Helbing, D. Brockmann, T. Chadefaux, K. Donnay, U. Blanke, O. Woolley, M. Moussaid, A. Johansson, J. Krause, S. Schutte, M. Perc, Saving human lives: what complexity science and information systems can contribute. J. Stat. Phys. **158**(3), 735–781 (2015)
5. D. Leite, P. Costa, F. Gomide, Evolving granular neural networks from fuzzy data streams. Neural Networks **38**, 1–16 (2013)
6. Y. Li, Q. Liu, S.-R. Tan, R.-H.-M. Chan, High-resolution time-frequency analysis of EEG signals using multiscale radial basis functions. Neurocomputing **195**, 96–103 (2016)
7. E. Lughofer, M. Sayed-Mouchaweh, Autonomous data stream clustering implementing split-and-merge concepts-towards a plug-and-play approach. Inf. Sci. **304**, 54–79 (2015)
8. E. Lughofer, Hybrid active learning for reducing the annotation effort of operators in classification systems. Pattern Recognit. **45**, 884–896 (2012)
9. C.J.B. Macnab, Using RBFs in a CMAC to prevent parameter drift in adaptive control. Neurocomputing **205**, 45–52 (2016)
10. M. Perc, A. Szolnoki, Coevolutionary games - a mini review. BioSystems **99**, 109–125 (2010)
11. H. Toubakh, M. Sayed-Mouchaweh, Hybrid dynamic classifier for drift-like fault diagnosis in a class of hybrid dynamic systems: application to wind turbine converters. Neurocomputing **171**, 1496–1516 (2016)
12. Q. Wu, X. Wang, Q. Shen, Research on dynamic modeling and simulation of axial-flow pumping system based on RBF neural network. Neurocomputing **186**, 200–206 (2016)
13. R. Yang, P.V. Er, Z. Wang, K.K. Tan, An RBF neural network approach towards precision motion system with selective sensor fusion. Neurocomputing **199**, 31–39 (2016)
14. C. Zhang, H. Wei, L. Xie, Y. Shen, K. Zhang, Direct interval forecasting of winds peed using radial basis function neural networks in a multi-objective optimization framework. Neurocomputing **205**, 53–63 (2016)
15. J. Cheng-Lv, Z. Yi, Y. Li, Non-divergence of stochastic discrete time algorithms for PCA neural networks. IEEE Trans. Neural Networks Learn. Syst. **26**(2), 394–399 (2015)
16. X. Li, R. Rakkiyappan, Stability results for Takagi-Sugeno fuzzy uncertain bam neural networks with time delays in the leakage term. Neural Comput. Appl. **22**(Supplement 1), S203-S219 (2013)
17. E. Lughofer, *Evolving Fuzzy Systems - Methodologies, Advanced Concepts and Applications* (Springer, Berlin, 2011). ISBN: 978-3-642-18086-6
18. W. Orozco-Tupacyupanqui, M. Nakano-Miyatake, H. Perez-Meana, A novel neural-fuzzy method to search the optimal step size for NLMS beamforming. IEEE Lat. Am. Trans. **13**(2), 402–408 (2015)
19. J.-J. Rubio, P. Angelov, J. Pacheco, An uniformly stable backpropagation algorithm to train a feedforward neural network. IEEE Trans. Neural Networks **22**(3), 356–366 (2011)
20. C.K. Ahn, A new solution to the induced l∞ finite impulse response filtering problem based on two matrix inequalities. Int. J. Control **87**(2), 404–409 (2014)

21. C.K. Ahn, An error passivation approach to filtering for switched neural networks with noise disturbance. Neural Comput. Appl. **21**(5), 853–861 (2012)
22. T. Hernandez-Cortes, J.-A. Meda-Campaña, L.-A. Paramo-Carranza, J.-C. Gomez-Mancilla, A simplified output regulator for a class of Takagi-Sugeno fuzzy models. Math. Prob. Eng. **2015**, 1–18 (2015)
23. J.A. Meda-Campaña, J. Rodriguez-Valdez, T. Hernandez-Cortes, R. Tapia-Herrera, V. Nosov, Analysis of the fuzzy controllability property and stabilization for a class of T-S fuzzy models. IEEE Trans. Fuzzy Syst. **23**(2), 291–301 (2015)
24. V. Nosov, J.-A. Meda-Campaña, J.-C. Gomez-Mancilla, J.-O. Escobedo-Alva, R.-G. Hernandez-Garcia, Stability analysis for autonomous dynamical switched systems through nonconventional Lyapunov functions. Math. Prob. Eng. **2015**, 1–18 (2015)
25. L. Zhang, Y. Zhu, W.X. Zheng, Energy-to-peak state estimation for Markov jump RNNs with time-varying delays via nonsynchronous filter with nonstationary mode transitions. IEEE Trans. Neural Networks Learn. Syst. **26**(10), 2346–2356 (2015)
26. L. Zhang, Y. Zhu, W.X. Zheng, Synchronization and state estimation of a class of hierarchical hybrid neural networks with time-varying delays. IEEE Trans. Neural Networks Learn. Syst. **27**(2), 459–470 (2016)
27. W. Gong, J. Liang, J. Cao, Matrix measure method for global exponential stability of complex-valued recurrent neural networks with time-varying delays. Neural Networks **70**, 81–89 (2015)
28. W. Gong, J. Liang, J. Cao, Global μ-stability of complex-valued delayed neural networks with leakage delay. Neurocomputing **168**, 135–144 (2015)
29. J. Qi, C. Li, T. Huang, Stability of inertial bam neural network with time-varying delay via impulsive control. Neurocomputing **161**, 162–167 (2015)
30. J.L. Wang, H.N. Wu, T. Huang, Passivity-based synchronization of a class of complex dynamical networks with time-varying delay. Automatica **56**, 105–112 (2015)
31. J.L. Wang, H.N. Wu, T. Huang, S.Y. Ren, J. Wu, Pinning control for synchronization of coupled reaction-diffusion neural networks with directed topologies. IEEE Trans. Syst. Man Cybern.: Syst. **46**(8), 1109–1120 (2016)
32. X. Wang, J. Yu, C. Li, H. Wang, T. Huang, J. Huang, Robust stability of stochastic fuzzy delayed neural networks with impulsive time window. Neural Networks **67**, 84–91 (2015)
33. J.J. Rubio, I. Elias, D.R. Cruz, J. Pacheco, Uniform stable radial basis function neural network for the prediction in two mechatronic processes. Neurocomputing **227**, 122–130 (2017)

Chapter 6
USNFIS: Uniform Stable Neuro Fuzzy Inference System

1 Introduction

The neuro fuzzy intelligent systems are the combination of the neural networks and the fuzzy systems which are applied for the learning of nonlinear behaviors. Some interesting investigations are detailed as follows. In [1, 2], new clustering algorithms utilized in the fault detection are proposed, in [3, 4], novel algorithms employed in the classification are described, and in [5, 6], metacognitive learning algorithms are introduced. Few researches have been carried out in the past to introduce this kind of algorithms to be utilized for the big data learning. Therefore, new efforts to increase the knowledge in this interesting issue would be of great interest.

Big data learning is the learning ability to solve via intelligent systems the problems where huge amounts of data are generated and updated during a short time; in this kind of systems the processing and analysis of data are important challenges. Some interesting works of this topic are described as follows. In [7–10], the classification of big data is focused, in [11–13], the modeling of big data is analyzed, and in [14, 15], the pattern recognition of big data is studied. In this chapter, there are two kinds of big data issues described as follows: (1) the systems with many inputs and outputs such as the considered in [7–9, 12, 14, 15] or (2) the systems with high changing data during a short time such as the considered in [10, 11, 13]; it is because in both cases huge amounts of data are generated and updated during a short time. This study is focused in case (2). On the other hand, in the aforementioned research, the stability of the algorithms is not analyzed, and the stability of the algorithms should be assured to avoid the damage of the devices due to the processing of big quantity of data.

The stable intelligent systems are the algorithms where the inputs, outputs, and parameters remain bounded through the time and where the overfitting is avoided. An algorithm with overfit has many parameters relative to the number of data; therefore, it has poor learning performance because it overreacts to minor fluctuations in the data. There is some research about the stable intelligent systems.

J. de J. Rubio, *Stability Analysis of Neural Networks and Evolving Intelligent Systems*, https://doi.org/10.1007/978-3-031-87282-2_6

In [16–18], the stability of continuous-time fuzzy neural networks is studied, in [19–22] the stability of the gradient algorithm is introduced, in [23–25], the stability of continuous-time neural networks is described, in [26, 27], the stability of discrete-time fuzzy systems is analyzed, in [28–30], the stability of continuous-time control systems is assured, and in [31, 32], the stability of controlled robotic systems is guaranteed. The stability should be assured in big data learning to guarantee a satisfactory behavior through the time for this kind of systems.

Most of the stable algorithms use a time varying learning speed such as the mentioned in [19, 21], and [33] for the learning of a multilayer neural network, or the mentioned in [20] and [22] for the learning of a fuzzy inference system. It is an efficient algorithm; therefore, it would be interesting to modify this algorithm to be applied in a neuro fuzzy system.

In this chapter, a neuro fuzzy inference system with a structure different with the multilayer neural network and the fuzzy inference system is suggested. Three differences between the introduced algorithm and the multilayer neural network and fuzzy inference system are described in function of the compactness, effectiveness, and stability as follows.

1. The suggested algorithm is different with the fuzzy inference system because the first only uses the numerator of the average defuzzifier while the second utilizes the average defuzzifier. The numerator of the average defuzzifier is better for the learning in big data than the average defuzzifier because the first is more compact than the second. Consequently, the suggested algorithm is compact.
2. The introduced algorithm is different with the multilayer neural network because the first employs Gaussian functions while the second utilizes sigmoid functions. The Gaussian function can be adapted better to the changing behavior of the systems than the sigmoid function because the first has three kinds of parameters while the second only uses two kinds of parameters and because the first considers positive and negative values, while the second only considers positive values. Therefore, the proposed algorithm is effective.
3. The proposed algorithm is different with both the multilayer neural network and fuzzy inference system because the first uses a time varying learning speed while the other uses a constant learning speed. The time varying learning speed is better for the learning in big data than the constant learning speed because the first reaches the stability and the boundedness of the parameters while the other does not. Thus, the introduced algorithm is stable.

The chapter is organized as follows. In Sect. 2, the neuro fuzzy inference system is presented. In Sect. 3, the closed loop dynamics of the neuro fuzzy inference system and the nonlinear system are obtained. In Sect. 4, the introduced algorithm for the big data learning of the neuro fuzzy inference system is designed. In Sect. 5, the stability, convergence, and boundedness of parameters for the aforementioned technique are guaranteed. In Sect. 6, the suggested strategy is summarized. In Sect. 7, the recommended algorithm is compared with other two algorithms for two processes. Section 8 presents the conclusions and suggests the future research directions.

2 Neuro Fuzzy Inference System

In this section, first, the big data nonlinear systems studied in this chapter are described, and second, the neuro fuzzy inference system for the big data learning of the nonlinear system behavior is introduced.

Consider the big data unknown discrete-time multiple input multiple output nonlinear system as follows:

$$y_{l*}(k) = f_l\left[Z_k\right], \tag{6.1}$$

where $i = 1, \ldots, N$, $l = 1, \ldots, O$, $Z_k = [z_1(k) \ldots, z_i(k), \ldots, z_N(k)]^T \in \Re^{N\times 1}$ is the input vector, N is the input number, O is the output number, $z_i(k) \in \Re$ and $y_{l*}(k) \in \Re$ are the inputs and outputs of the plant, and f is an unknown and smooth nonlinear function, $f_l \in C^{\infty}$.

The neuro fuzzy inference system with one hidden layer for the big data learning of the nonlinear system (6.1) is described as follows:

$$\begin{aligned} y_l(k) = d_l(k) &= \sum_{j=1}^{M} a_{jl}(k)\alpha_j(u_j(k)), \\ \alpha_j(u_j(k)) &= e^{-u_j^2(k)}, \\ u_j(k) &= \sum_{i=1}^{N} b_{ij}(k)\left[z_i(k) - c_i(k)\right], \end{aligned} \tag{6.2}$$

where $i = 1, \ldots, N$, $j = 1, \ldots, M$, $l = 1, \ldots, O$, $z_i(k) \in \Re$ and $y_l(k) \in \Re$ are the inputs and outputs of the neuro fuzzy inference system, $a_{jl}(k) \in \Re$, $b_{ij}(k) \in \Re$, $c_i(k) \in \Re$ are the parameters of the output layer, hidden layer, and centers, $\alpha_j(u_j(k)) \in \Re$ is a nonlinear function, $u_j(k) \in \Re$ is the addition function, M is the number of neurons in the hidden layer, and O is the output number. Figure 6.1 shows the architecture of the neuro fuzzy inference system where the input layer, hidden layer, and output layer are observed.

Remark 6.1 In [11, 34–36], and [37], the interesting radial basis function neural networks are considered. A radial basis function neural network cannot be seen as a multilayer neural network because the first utilizes the Gaussian functions while the second employs the sigmoid functions. From [38, 39], a radial basis function neural network can be seen as a fuzzy inference system because both use the Gaussian functions.

Remark 6.2 The fuzzy inference system is given as follows:

$$[c]cy_l(k) = d_l(k) = \frac{\sum_{j=1}^{M} a_{jl}(k)\alpha_j(u_j(k))}{\sum_{j=1}^{M} \alpha_j(u_j(k))}, \tag{6.3}$$

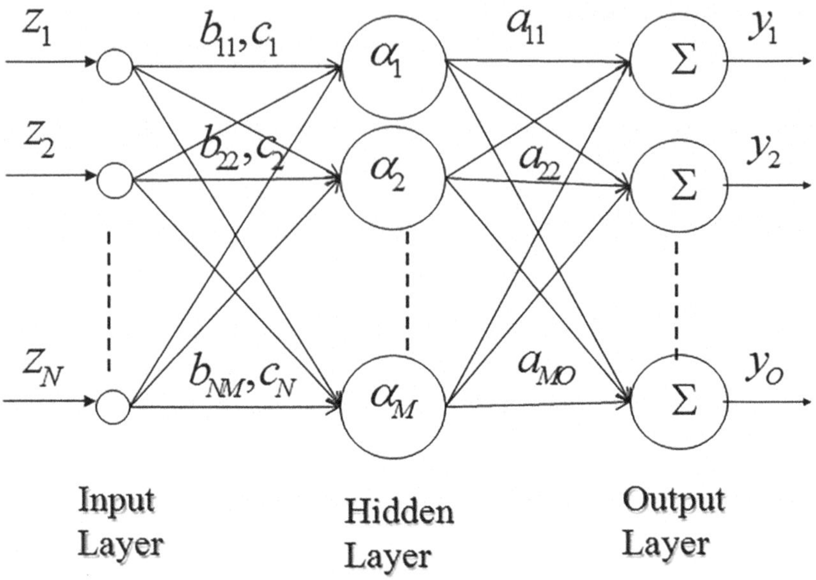

Fig. 6.1 Architecture of the neuro fuzzy inference system

$$\alpha_j(u_j(k)) = e^{-u_j^2(k)},$$

$$u_j(k) = \sum_{i=1}^{N} b_{ij}(k)\left[z_i(k) - c_i(k)\right],$$

And the numerator of the average defuzzifier described by the first equation of (6.2) is more compact than the average defuzzifier described by the first equation of (6.3) because the first utilizes a less number of operations than the second.

Remark 6.3 The multilayer neural network is given as follows:

$$\begin{aligned} y_l(k) = d_l(k) &= \sum_{j=1}^{M} a_{jl}(k)\alpha_j(u_j(k)), \\ \alpha_j(u_j(k)) &= sig\left[u_j(k)\right], \\ u_j(k) &= \sum_{i=1}^{N} b_{ij}(k) z_i(k); \end{aligned} \tag{6.4}$$

the Gaussian function explained by the second and third equations of (6.2) is more effective than the sigmoid function explained by the second and third equations of (6.4) because the first utilizes two kinds of parameters while the second employs only one kind of parameter.

3 Closed Loop Dynamics of the Neuro Fuzzy Inference System

In this section, the closed loop dynamics of the neuro fuzzy system applied for the big data learning of the nonlinear system behavior are obtained via the linearization technique. The closed loop dynamics of the neuro fuzzy inference system are required for the algorithm design which described in the next section and for the stability analysis which is explained two sections later.

According to the Stone-Weierstrass theorem [13, 21], the unknown nonlinear function f of (6.1) is approximated as follows:

$$\begin{aligned} y_{l*}(k) &= d_{l*} + \in_{lf} = \sum_{j=1}^{M} a_{jl*}\alpha_{j*} + \in_{lf}, \\ \alpha_{j*} &= e^{-u_{j*}^{2}}, \\ u_{j*} &= \sum_{i=1}^{N} b_{ij*}\left[z_i(k) - c_{i*}\right], \end{aligned} \tag{6.5}$$

where $\in_{lf} = y_{l*}(k) - d_{l*} \in \Re$ is the modeling error, $\alpha_{j*} \in \Re$, $a_{jl*} \in \Re$, $b_{ij*} \in \Re$, and $c_{i*} \in \Re$ are the optimal parameters that can minimize the modeling error $\in_{lf}$. In the case of three independent variables, a function has a Taylor series as follows:

$$\begin{aligned} f_l(\omega_1, \omega_2, \omega_3) &= f_l(\omega_{1^0}, \omega_{2^0}, \omega_{3^0}) + \left(\omega_1 - \omega_{1^0}\right)\frac{\partial f_l(\omega_1,\omega_2,\omega_3)}{\partial \omega_1} \\ &+ \left(\omega_2 - \omega_{2^0}\right)\frac{\partial f_l(\omega_1,\omega_2,\omega_3)}{\partial \omega_2} + \left(\omega_3 - \omega_{3^0}\right)\frac{\partial f_l(\omega_1,\omega_2,\omega_3)}{\partial \omega_3} + \xi_{lf}, \end{aligned} \tag{6.6}$$

where $\xi_{lf} \in \Re$ is the remainder of the Taylor series. ω_1, ω_2, and ω_3 correspond to $a_{jl}(k) \in \Re$, $b_{ij}(k) \in \Re$, and $c_i(k) \in \Re$, ω_{1^0}, ω_{2^0}, and ω_{3^0} correspond to $a_{jl*} \in \Re$, $b_{ij*} \in \Re$, and $c_{i*} \in \Re$, define $\widetilde{a}_{jl}(k) = a_{jl}(k) - a_{jl*} \in \Re$, $\widetilde{b}_{ij}(k) = b_{ij}(k) - b_{ij*} \in \Re$, and $\widetilde{c}_i(k) = c_i(k) - c_{i*} \in \Re$; therefore, the Taylor series is applied to obtain the closed loop dynamics of the neuro fuzzy inference system (6.2) and the nonlinear system behavior (6.1) as follows:

$$\begin{aligned} d_l(k) &= d_{l*} + \sum_{j=1}^{M} \widetilde{a}_{jl}(k)\frac{\partial d_l(k)}{\partial a_{jl}(k)} + \sum_{i=1}^{N}\sum_{j=1}^{M} \widetilde{b}_{ij}(k)\frac{\partial d_l(k)}{\partial b_{ij}(k)} \\ &+ \sum_{i=1}^{N} \widetilde{c}_i(k)\frac{\partial d_l(k)}{\partial c_i(k)} + \xi_{lf}, \end{aligned} \tag{6.7}$$

where $\frac{\partial d_l(k)}{\partial a_{jl}(k)} \in \Re$, $\frac{\partial d_l(k)}{\partial b_{ij}(k)} \in \Re$, and $\frac{\partial d_l(k)}{\partial c_i(k)} \in \Re$; please note that $d_l(k) = \sum_{j=1}^{M} a_{jl}(k)\alpha_j(u_j(k)) \in \Re$. Since all the parameters are scalars, the Taylor series is fully applicable. Considering (6.2) and using the chain rule, it gives

$$\frac{\partial d_l(k)}{\partial a_{jl}(k)} = F_j(k) = \alpha_j(u_j(k)), \tag{6.8}$$

where $\alpha_j(u_j(k)) = e^{-u_j^2(k)}$ and $u_j(k)$ are given in (6.2). Utilizing the same process gives

$$\begin{aligned} \tfrac{\partial d_l(k)}{\partial b_{ij}(k)} &= G_{ijl}(k) \\ &= \gamma_j(u_j(k))a_{jl}(k)\left[c_i(k) - z_i(k)\right], \end{aligned} \tag{6.9}$$

where $\gamma_j(u_j(k)) = 2u_j(k)\alpha_j(u_j(k)) \in \Re$. Again employing the same process gives

$$\begin{aligned} \tfrac{\partial d_l(k)}{\partial c_i(k)} &= H_{il}(k) \\ &= b_{ij}(k)\gamma_j(u_j(k))a_{jl}(k), \end{aligned} \tag{6.10}$$

Substituting $\frac{\partial d_l(k)}{\partial a_{jl}(k)}$ of (6.8), $\frac{\partial d_l(k)}{\partial b_{ij}(k)}$ of (6.9), and $\frac{\partial d_l(k)}{\partial c_i(k)}$ of (6.10) into (6.7), it gives

$$\begin{aligned} d_l(k) = d_{l*} &+ \sum_{j=1}^{M} \widetilde{a}_{jl}(k)F_j(k) + \sum_{i=1}^{N}\sum_{j=1}^{M} \widetilde{b}_{ij}(k)G_{ijl}(k) \\ &+ \sum_{i=1}^{N} \widetilde{c}_i(k)H_{il}(k) + \xi_{lf}. \end{aligned} \tag{6.11}$$

Define the learning error $\widetilde{y}_l(k) \in \Re$ as follows:

$$\widetilde{y}_l(k) = y_l(k) - y_{l*}(k), \tag{6.12}$$

where $y_{l*}(k)$ and $y_l(k)$ are defined in (6.1) and (6.2), respectively. Substituting (6.2), (6.5), and (6.12) into (6.11) gives closed loop dynamics:

$$\begin{aligned} \widetilde{y}_l(k) = &\sum_{j=1}^{M} \widetilde{a}_{jl}(k)F_j(k) + \sum_{i=1}^{N}\sum_{j=1}^{M} \widetilde{b}_{ij}(k)G_{ijl}(k) \\ &+ \sum_{i=1}^{N} \widetilde{c}_i(k)H_{il}(k) + \mu_l(k), \end{aligned} \tag{6.13}$$

where $\mu_l(k) = \xi_{lf} - \epsilon_{lf}$.

4 Design of the Recommended Algorithm

In this section, the recommended algorithm utilized in a neuro fuzzy inference system is designed for the big data learning of the nonlinear system behavior. In this part, the adapting law of the proposed algorithm is obtained.

Theorem 6.1 *The introduced algorithm that is the updating function of the neuro fuzzy inference system (6.2) for the big data learning of the nonlinear system (6.1) is given as follows:*

$$\begin{aligned} a_{jl}(k+1) &= a_{jl}(k) - \eta(k)F_j(k)\widetilde{y}_l(k), \\ b_{ij}(k+1) &= b_{ij}(k) - \eta(k)G_{ijl}(k)\widetilde{y}_l(k), \\ c_i(k+1) &= c_i(k) - \eta(k)H_{il}(k)\widetilde{y}_l(k), \end{aligned} \tag{6.14}$$

where $F_j(k)$, $G_{ijl}(k)$, and $H_{il}(k)$ are given in (6.8), (6.9), and (6.10), respectively, and $\widetilde{y}_l(k)$ is the learning error of (6.12).

Proof *See [40] for the proof.*

5 Stability Analysis of the Introduced Algorithm

In this section, the recommended stable algorithm utilized in a neuro fuzzy inference system is designed for the big data learning of the nonlinear system behavior. In this part, the time varying learning speed used in the adapting law of the proposed algorithm is suggested; furthermore, the stability and convergence of the introduced algorithm are assured.

The introduced algorithm is given in (6.14) with a time varying learning speed as follows:

$$\begin{aligned} a_{jl}(k+1) &= a_{jl}(k) - \eta(k)F_j(k)\widetilde{y}_l(k), \\ b_{ij}(k+1) &= b_{ij}(k) - \eta(k)G_{ijl}(k)\widetilde{y}_l(k), \\ c_i(k+1) &= c_i(k) - \eta(k)H_{il}(k)\widetilde{y}_l(k), \end{aligned} \tag{6.15}$$

where the new time varying learning speed $\eta(k)$ is as follows:

$$\eta(k) = \frac{\eta_0}{2\left(\frac{1}{2} + \sum_{j=1}^{M} F_j^2(k) + \sum_{i=1}^{N}\sum_{j=1}^{M} G_{ijl}^2(k) + \sum_{i=1}^{N} H_{il}^2(k)\right)},$$

where $i = 1, \ldots, N$, $j = 1, \ldots, M$, $l = 1, \ldots, O$, $F_j(k) \in \Re$ is defined in (6.8), $G_{ijl}(k) \in \Re$ is defined in (6.9), $H_{il}(k) \in \Re$ is defined in (6.10), $\widetilde{y}_l(k)$ is defined in (6.12), $0 < \eta_0 \leq 1 \in \Re$, consequently $0 < \eta(k) \in \Re$, $\overline{\mu}_l$ is its upper bound of the uncertainty $\mu_l(k)$, $|\mu_l(k)| < \overline{\mu}_l$.

Remark 6.4 $\eta(k)$ is the one main part of the recommended algorithm, and it is selected by the designer as an average and bounded function such as the stability of the algorithm (6.15) can be assured.

The following theorem gives the stability of the suggested algorithm.

Theorem 6.2 *The algorithm (6.2), (6.12), and (6.15) applied for the big data learning of the nonlinear system (6.1) is uniformly stable, and the upper bound of the average learning error* $\widetilde{y}_{lp}^2(k)$ *satisfies*

$$\limsup_{T\to\infty} \frac{1}{T}\sum_{k=2}^{T} \widetilde{y}_{lp}^2(k) \leq \alpha_0 \overline{\mu}_l^2, \tag{6.16}$$

where $\widetilde{y}_{lp}^2(k) = \frac{\eta(k-1)}{2}\widetilde{y}_l^2(k-1)$, $0 < \eta_0 \leq 1 \in \Re$ *and* $0 < \eta(k) \in \Re$ *are defined in (6.15),* $\widetilde{y}_l(k)$ *is defined in (6.12),* $\overline{\mu}_l$ *is the upper bound of the uncertainty* $\mu_l(k)$, $|\mu_l(k)| < \overline{\mu}_l$.

Proof See [40] for the proof.

Remark 6.5 There are two requirements to apply this algorithm for the big data learning of the nonlinear system behavior: the first is that the uncertainty $\mu_l(k)$ should be bounded, and the second is that the nonlinear system should have the structure described by Eq. (6.1).

Remark 6.6 The bound of $\mu_l(k)$ denoted as $\overline{\mu}_l$ is not utilized in the suggested algorithm (6.2), (6.12), (6.15) because it is only considered to assure its stability.

The following theorem proves that the parameters of the introduced algorithm are bounded.

Theorem 6.3 *When the average learning error* $\widetilde{y}_{lp}^2(k+1)$ *is bigger than the uncertainty* $\eta_0\overline{\mu}_l^2$*, the parameters error is bounded by the initial parameters error as follows:*

$$\begin{gathered} \widetilde{y}_{lp}^2(k+1) \geq \eta_0 \overline{\mu}^2 \\ \Longrightarrow \sum_{j=1}^{M} \widetilde{a}_{jl}^2(k) + \sum_{j=1}^{M}\sum_{i=1}^{N} \widetilde{b}_{ij}^2(k) + \sum_{i=1}^{N} \widetilde{c}_i^2(k) \\ \leq \sum_{j=1}^{M} \widetilde{a}_{jl}^2(1) + \sum_{j=1}^{M}\sum_{i=1}^{N} \widetilde{b}_{ij}^2(1) + \sum_{i=1}^{N} \widetilde{c}_i^2(1), \end{gathered} \tag{6.17}$$

where $i = 1, \ldots, N$, $j = 1, \ldots, M$, $l = 1, \ldots, O$, $\widetilde{a}_{jl}(k)$, $\widetilde{b}_{ij}(k)$, *and* $\widetilde{c}_i(k)$ *are defined in (6.6),* $\widetilde{a}_{jl}(1)$, $\widetilde{b}_{ij}(1)$, *and* $\widetilde{c}_i(1)$ *are the initial parameters errors,* $\widetilde{y}_{lp}^2(k+$

1) $= \frac{1}{2}\eta(k)\widetilde{y}_l^2(k)$, $a_{jl}(k+1)$, $b_{ij}(k+1)$, $c_i(k+1)$, $0 < \eta_0 \leq 1 \in \Re$, and $0 < \eta(k) \in \Re$ are defined in (6.15), $\widetilde{y}_l(k)$ is defined in (6.12), $\overline{\mu}_l$ is the upper bound of the uncertainty $\mu_l(k)$, $|\mu_l(k)| < \overline{\mu}_l$.

Proof *See [40] for the proof.*

Remark 6.7 From Theorem 6.2 the average learning error $\widetilde{y}_{lp}^2(k+1)$ of the suggested method is bounded, and from Theorem 6.3 the parameters errors $\widetilde{a}_{jl}^2(k)$, $\widetilde{b}_{ij}^2(k)$, and $\widetilde{c}_i^2(k)$ are bounded, i.e., the introduced technique for the learning of a neuro fuzzy inference system is uniformly stable in the presence of unmodeled dynamics, and the overfitting is avoided. Furthermore, the learning error converges to a small zone bounded by the unmodeled dynamics $\overline{\mu}_l$.

6 The Suggested Algorithm

In this section, the steps of the application for suggested algorithm are explained.

(1) Obtain the outputs of the nonlinear system $y_{l*}(k)$ with Eq. (6.1). Note that the nonlinear system may have the structure represented by Eq. (6.1); the parameters N and O are selected according to the input and output number of this nonlinear system.
(2) Select the following parameters: $a_{jl}(1)$, $b_{ij}(1)$, and $c_i(1)$ as random numbers between 0 and 1, M as an integer number, and η_0 as a positive value smaller than or equal to 1; obtain the outputs of the neuro fuzzy inference system $y_l(1)$ with Eq. (6.2).
(3) For each iteration k, obtain the outputs of the neuro fuzzy inference system $y_l(k)$ with Eq. (6.2), also obtain the learning error $\widetilde{y}_l(k)$ with Eq. (6.12), and update the parameters $a_{jl}(k+1)$, $b_{ij}(k+1)$, and $c_i(k+1)$ with Eq. (6.15).
(4) Note that the behavior of the algorithm could be improved by selecting other values for M or η_0.

Remark 6.8 The focused neuro fuzzy inference system has one hidden layer. A neuro fuzzy inference system with one hidden layer is sufficient to approximate any nonlinear system.

7 Results

In this section, two examples are considered. In the examples, the suggested algorithm is applied for the big data learning of the crude oil blending process and the beetle population process. In all cases, the recommended algorithm denoted as USNFIS is compared with the fuzzy inference system of [22] denoted as FIS and with the gradient algorithm of [21] denoted as G. It is important to note that the three mentioned algorithms are stable with the difference that the G is a multilayer neural network which utilizes sigmoid functions, the FIS is a fuzzy inference system which employs the average defuzzifier, while the USNFIS is a uniform stable fuzzy inference system which considers Gaussian functions and the numerator of the average defuzzifier. The root mean square error (RMSE) is used for the algorithms comparison, and it is given as follows:

$$\mathrm{RMSE} = \left(\frac{1}{T} \sum_{k=1}^{T} \sum_{l=1}^{O} \widetilde{y}_l^2(k) \right)^{\frac{1}{2}}, \tag{6.18}$$

where $\widetilde{y}_l(k)$ is the learning error of (6.12), T is the iteration number, and O is the output number.

7.1 Crude Oil Blending Process

In this example, the studied algorithm is applied for the modeling of the crude oil blending process [13]. One neuro fuzzy inference system is utilized for the training, and the same system is utilized for the testing. The crude oil blending process has six inputs and three outputs which are high changing data during a short time. The inputs are $z_1(k) = L_3$, $z_2(k) = Puerto\ Ceiba$, and the output is $y_{1*}(k) = Q_a$ for the first blending process, the inputs are $z_3(k) = Q_b$, $z_4(k) = Maya$, and the output is $y_{2*}(k) = Q_c$ for the second blending process, and the inputs are $z_5(k) = Q_c$, $z_6(k) = El\ Golpe$, and the output is $y_{3*}(k) = I_n = International$ for the third blending process. In all the cases the ^{o}API is considered. The data of 7875 iterations of operation are used for the training, and the data of the least 525 iterations are used for the testing.

G is given by Rubio et al. [21] with parameters $N = 6$, $O = 3$, $M = 10$, $\alpha_0 = 1$, and V_{j1} and W_{ij1} are random numbers between 0 and 1.

FIS is given by Rubio [22] with parameters $N = 6$, $O = 3$, $M = 10$, $\eta_0 = 1$, and $a_{jl}(1)$, $b_{ij}(1)$, and $c_i(1)$ are random numbers between 0 and 1.

USNFIS is given as (6.2), (6.12), and (6.15) with parameters $N = 6$, $O = 3$, $M = 10$, $\eta_0 = 1$, $a_{jl}(1)$, $b_{ij}(1)$, and $c_i(1)$ are random numbers between 0 and 1.

Figure 6.2 shows the comparison results for the average learning error where in USNFIS the final average error is 7.6086×10^{-4}, in FIS of [22] the final average error is 0.0020, and in G of [21] the final average error is 0.0016. Figure 6.3 shows the training results, and Fig. 6.4 shows the testing results. Table 6.1 shows the training RMSE results, and Table 6.2 shows the testing RMSE results for many intermediate iterations termed with th via Eq. (6.18).

From Figs. 6.2, 6.3, and 6.4, it is observed that USNFIS is better than both the G and FIS because the signal of the first follows better the signal of the plant than the signal of the other. From Tables 6.1 and 6.2, it can be observed that the USNFIS obtained better accuracy when it is compared with both G and FIS because

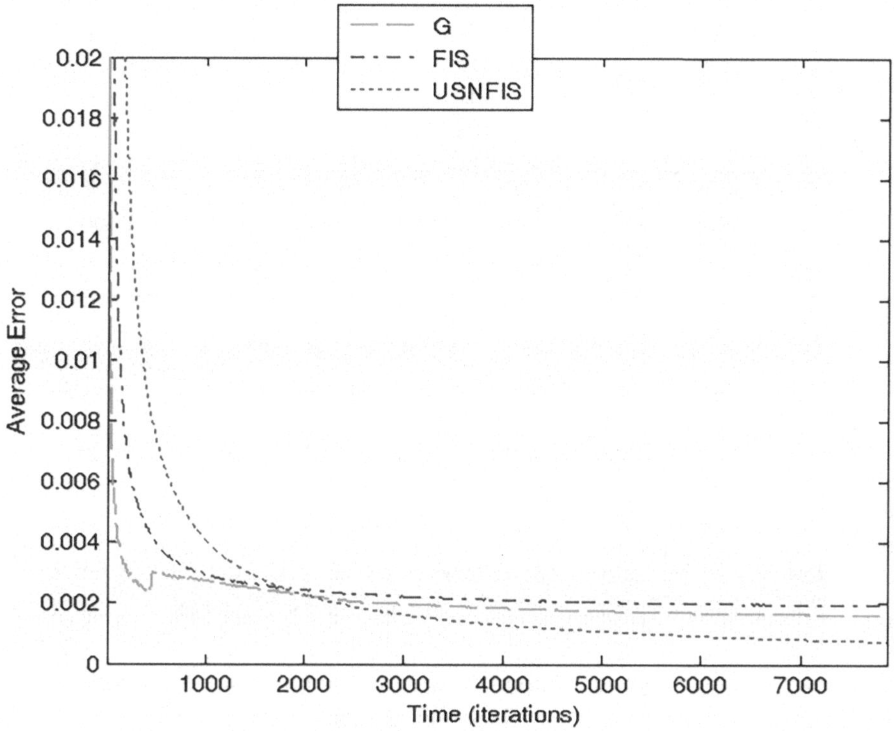

Fig. 6.2 Average learning errors for the oil blending process

the RMSE is smaller for the first. Thus, the USNFIS is preferable for the big data learning of the oil blending process.

7.2 *Beetle Population Process*

In this example, the introduced algorithm is applied for the modeling on the flour beetle population [41]. The beetle population process has six inputs and three outputs which are high changing data during a short time. The model of experimental population studies of a model of flour beetle population dynamics describes an age-structured population:

$$\begin{aligned} L(k+1) &= b_f A(k) e^{-c_{ea} A(k) - c_{el} L(k)}, \\ P(k+1) &= [1 - \mu_l]\, L(k), \\ A(k+1) &= P(k) e^{-c_{pa} A(k)} + [1 - \mu_a]\, A(k), \end{aligned} \tag{6.19}$$

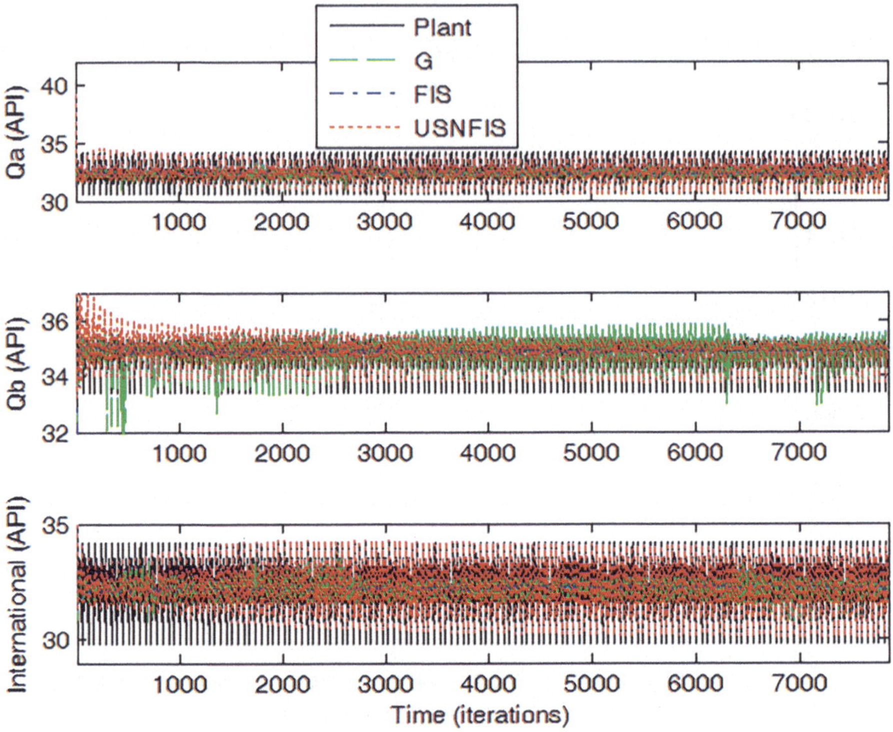

Fig. 6.3 Training results for the oil blending process

where:
b_f = Larvae recruits per adult = 11.98 numbers
c_{ea} = Susceptibility of eggs to cannibalism by adults = 0.011 unitless
c_{el} = Susceptibility of eggs to cannibalism by larvae = 0.013 unitless
c_{pa} = Susceptibility of pupae to cannibalism by adults = 0.017 unitless
μ_l = Fraction of larvae dying (not cannibalism) = 0.513 unitless
μ_a = Fraction of adults dying = 0.96 unitless

$L(k)$ are the Larvae which starts with 250 numbers, $P(k)$ are the Pupae which starts with 5 numbers, and $A(k)$ are the Adults which starts with 100 numbers. The data of 7800 iterations of operation are used for the training, and the data of the least 200 iterations are used for the testing. One neuro fuzzy inference system is used for the training, and the same system is used for the testing. $z_1(k) = P(k)$ and $z_2(k) = A(k)$ are the inputs, and $y_{1*}(k) = L(k+1)$ is the output for the training of the first population process. $z_3(k) = L(k)$ and $z_4(k) = A(k)$ are the inputs, and $y_{2*}(k) = P(k+1)$ is the output for the training of the second population process. Finally, $z_5(k) = L(k)$ and $z_6(k) = P(k)$ are the inputs, and $y_{3*}(k) = A(k+1)$ is the output for the training of the third population process.

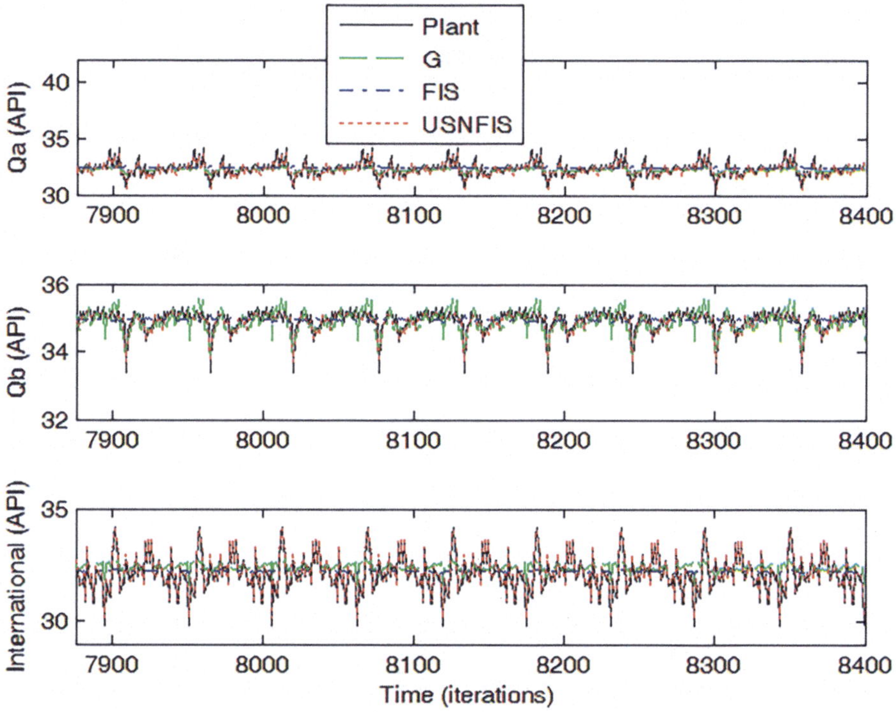

Fig. 6.4 Testing results for the oil blending process

Table 6.1 RMSE training results for the oil blending process

Techniques	1000th	2000th	3000th	4000th	5000th	6000th	7000th	7875th
G	0.1588	0.1473	0.1419	0.1409	0.1404	0.1399	0.1389	0.1389
FIS	0.1624	0.1580	0.1560	0.1547	0.1540	0.1534	0.1530	0.1528
USNFIS	0.1473	0.1183	0.1037	0.0946	0.0882	0.0833	0.0794	0.0766

Table 6.2 RMSE testing results for the oil blending process

Techniques	8075th	8275th	8400th
G	0.0353	0.0311	0.0228
FIS	0.0375	0.0331	0.0240
USNFIS	0.0122	0.0107	0.0077

G is given by Rubio et al. [21] with parameters $N = 6$, $O = 3$, $M = 10$, $\alpha_0 = 1$, and V_{j1} and W_{ij1} are random numbers between 0 and 1.

FIS is given by Rubio [22] with parameters $N = 6$, $O = 3$, $M = 10$, $\eta_0 = 1$, and $a_{jl}(1)$, $b_{ij}(1)$, and $c_i(1)$ are random numbers between 0 and 1.

USNFIS is given as (6.2), (6.12), and (6.15) with parameters $N = 6$, $O = 3$, $M = 10$, $\eta_0 = 1$, and $a_{jl}(1)$, $b_{ij}(1)$, and $c_i(1)$ are random numbers between 0 and 1.

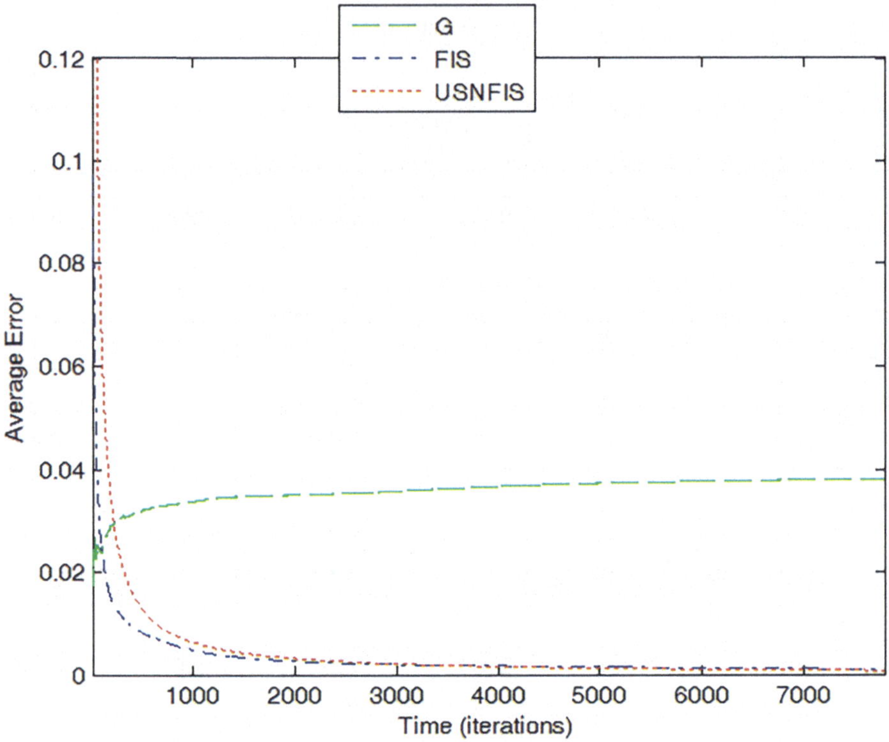

Fig. 6.5 Average learning errors for the beetle population process

Figure 6.5 shows the comparison results of the average learning error where in USNFIS the final average error is 8.5516×10^{-4}, in FIS of [22] the final average error is 0.0011, and in G of [21] the final average error is 0.0382. Figure 6.6 shows the training results, and Fig. 6.7 shows the testing results. Table 6.3 shows the training RMSE results, and Table 6.4 shows the testing RMSE results for many intermediate iterations termed with th via Eq. (6.18).

From Figs. 6.5, 6.6, and 6.7, it can be observed that the USNFIS is better than both G and FIS because the signal of the first follows better the signal of the plant than the signal of the other. From Tables 6.3 and 6.4, it is observed that the USNFIS achieves better accuracy when it is compared with both G and FIS because the RMSE is smaller for the first. Thus, the USNFIS is preferable for the big data learning of the beetle population process.

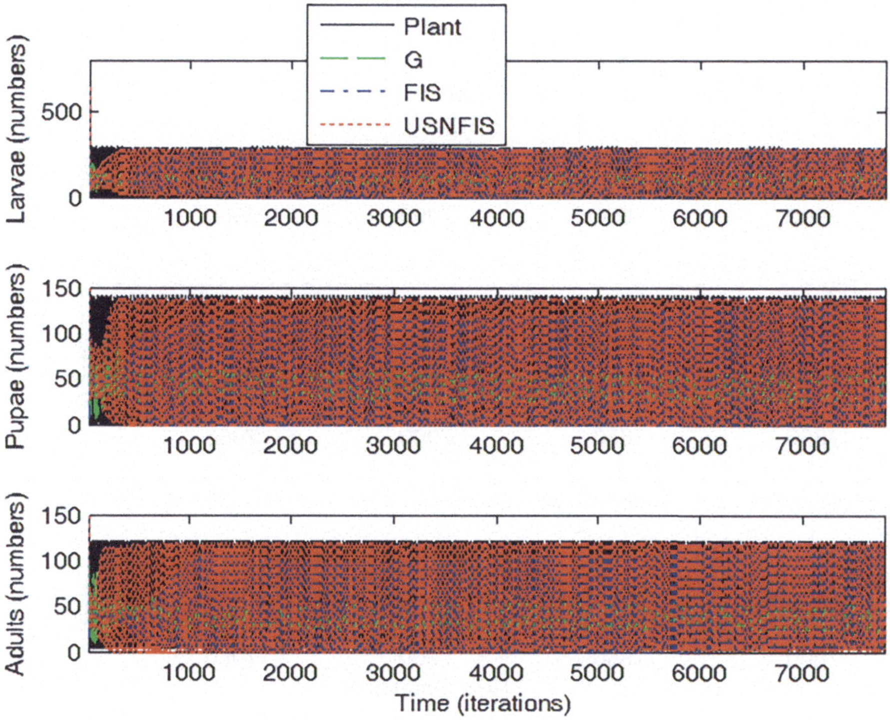

Fig. 6.6 Training results for the beetle population process

8 Concluding Remarks

In this chapter, a novel algorithm is designed for the big data learning of a neuro fuzzy inference system, and the stability, convergence, and boundedness of parameters for the studied technique are guaranteed. From the results, it is shown that the introduced approach achieves better accuracy for the big data learning of nonlinear system behaviors when it is compared with both the gradient and fuzzy inference system methods. The suggested technique could be used to train a neuro fuzzy inference system such as it is applied in this chapter, or it could be used as the parameters updating of an evolving intelligent system. As a future research, the focused method will be applied in the control or in the evolving intelligent systems, or other new algorithms will be designed.

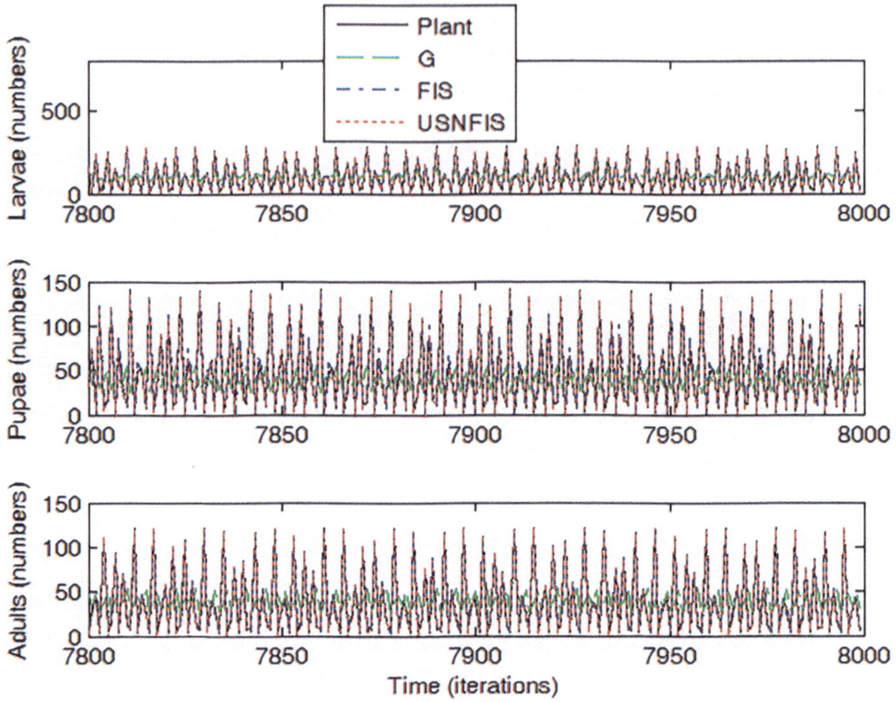

Fig. 6.7 Training results for the beetle population process

Table 6.3 RMSE training results for the beetle population process

Techniques	1000th	2000th	3000th	4000th	5000th	6000th	7000th	7800th
G	0.4267	0.4263	0.4257	0.4254	0.4252	0.4250	0.4248	0.4247
FIS	0.3092	0.2291	0.1967	0.1788	0.1652	0.1536	0.1437	0.1369
USNFIS	0.1841	0.1321	0.1089	0.0951	0.0856	0.0786	0.0732	0.0696

Table 6.4 RMSE testing results for the beetle population process

Techniques	7900th	8000th
G	0.0644	0.0460
FIS	0.0066	0.0047
USNFIS	0.0027	0.0019

References

1. E. Lughofer, M. Sayed-Mouchaweh, Autonomous data stream clustering implementing split-and-merge concepts-towards a plug-and-play approach. Inf. Sci. **304**, 54–79 (2015)
2. E. Lughofer, Hybrid active learning for reducing the annotation effort of operators in classification systems. Pattern Recognit. **45**, 884–896 (2012)

3. M. Pratama, S.G. Anavatti, E. Lughofer, GENEFIS: toward an effective localist network. IEEE Trans. Fuzzy Syst. **22**(3), 547–562 (2014)
4. M. Pratama, S.G. Anavatti, M.-J. Er, E. Lughofer, pClass: an effective classifier for streaming examples. IEEE Trans. Fuzzy Syst. **23**(2), 369–386 (2015)
5. M. Pratama, G. Zhang, M.J. Er, S. Anavatti, An incremental type-2 meta-cognitive extreme learning machine. IEEE Trans. Cybernet. **47**(2), 339–353 (2017)
6. M. Pratama, J. Lu, S. Anavatti, E. Lughofer, C.-P. Lim, An incremental meta-cognitive-based scaffolding fuzzy neural network. Neurocomputing **171**, 89–105 (2016)
7. D. Kangin, P. Angelov, J.A. Iglesias, A. Sanchis, Evolving classifier TEDAClass for big data. Procedia Comput. Sci. **53**, 9–18 (2015)
8. A. Roy, A classification algorithm for high-dimensional data. Procedia Comput. Sci. **53**, 345–355 (2015)
9. A. Roy, P.D. Mackin, S. Mukhopadhyay, Methods for pattern selection, class-specific feature selection and classification for automated learning. Neural Networks **41**, 113–129 (2013)
10. A. Roy, On findings of category and other concept cells in the brain: some theoretical perspectives on mental representation. Cognit. Comput. **7**, 279–284 (2015)
11. Y. Li, Q. Liu, S.R. Tan, R.H.M. Chan, High-resolution time-frequency analysis of EEG signals using multiscale radial basis functions. Neurocomputing **195**, 96–103 (2016)
12. B. Luitel, G.K. Venayagamoorthy, Cellular computational networks-a scalable architecture for learning the dynamics of large networked systems. Neural Networks **50**, 120–123 (2014)
13. J.J. Rubio, Least square neural network model of the crude oil blending process. Neural Networks **78**, 88–96 (2016)
14. B. Xu, K. Huang, I. King, C.L. Liu, J. Sun, N. Satoshi, Graphical lasso quadratic discriminant function and its application to character recognition. Neurocomputing **129**, 33–40 (2014)
15. M.C. Yuen, I. King, K.S. Leung, TaskRec: a task recommendation framework in crowdsourcing systems. Neural Proces. Lett. **41**(2), 223–238 (2015)
16. C.K. Ahn, A new solution to the induced l∞ finite impulse response filtering problem based on two matrix inequalities. Int. J. Control **87**(2), 404–409 (2014)
17. C.K. Ahn, M.T. Lim, Model predictive stabilizer for T-S fuzzy recurrent multilayer neural network models with general terminal weighting matrix. Neural Comput. Appl. **23**(suppl 1), S271-S277 (2013)
18. C.K. Ahn, An error passivation approach to filtering for switched neural networks with noise disturbance. Neural Comput. Appl. **21**(5), 853–861 (2012)
19. J. Cheng-Lv, Z. Yi, Y. Li, Non-divergence of stochastic discrete time algorithms for PCA neural networks. IEEE Trans. Neural Networks Learn. Syst. **26**(2), 394–399 (2015)
20. J.A. Meda-Campaña, J. Rodriguez-Valdez, T. Hernandez-Cortes, R. Tapia-Herrera, V. Nosov, Analysis of the fuzzy controllability property and stabilization for a class of T-S fuzzy models. IEEE Trans. Fuzzy Syst. **23**(2), 291–301 (2015)
21. J.J. Rubio, P. Angelov, J. Pacheco, An uniformly stable backpropagation algorithm to train a feedforward neural network. IEEE Trans. Neural Networks **22**(3), 356–366 (2011)
22. J.J. Rubio, Fuzzy slopes model of nonlinear systems with sparse data. Soft Comput. **19**(12), 3507–3514 (2015)
23. R. Rakkiyappan, R. Sasirekha, Y. Zhu, L. Zhang, H∞ state estimator design for discrete-time switched neural networks with multiple missing measurements and sojourn probabilities. J. Franklin Instit. **353**, 1358–1385 (2016)
24. L. Zhang, Y. Zhu, W.X. Zheng, Energy-to-peak state estimation for Markov jump RNNs with time-varying delays via nonsynchronous filter with nonstationary mode transitions. IEEE Trans. Neural Networks Learn. Syst. **26**(10), 2346–2356 (2015)
25. L. Zhang, Y. Zhu, W.X. Zheng, Synchronization and state estimation of a class of hierarchical hybrid neural networks with time-varying delays. IEEE Trans. Neural Networks Learn. Syst. **27**(2), 459–470 (2016)
26. Z. Ning, L. Zhang, J.J. Rubio, X. Yin, Asynchronous filtering for discrete-time fuzzy affine systems with variable quantization density. IEEE Trans. Cybernet. **47**(1), 153–164 (2017)

27. L. Zhang, T. Yang, P. Shi, M. Liu, Stability and stabilization of a class of discrete-time fuzzy systems with semi-Markov stochastic uncertainties. IEEE Trans. Syst. Man Cybern.: Syst. **46**(12), 1642–1653 (2016)
28. Y. Pan, M.J. Er, D. Huang, Q. Wang, Adaptive fuzzy control with guaranteed convergence of optimal approximation error. IEEE Trans. Fuzzy Syst. **19**(5), 807–818 (2011)
29. T. Sun, H. Pei, Y. Pan, C. Zhang, Robust wavelet network control for a class of autonomous vehicles to track environmental contour line. Neurocomputing **74**, 2886–2892 (2011)
30. Y. Pan, H. Yu, Biomimetic hybrid feedback feedforward neural-network learning control. IEEE Trans. Neural Networks Learn. Syst. **28**(6), 1481–1487 (2017)
31. Y. Pan, Y. Liu, B. Xu, H. Yu, Hybrid feedback feedforward: an efficient design of adaptive neural network control. Neural Networks **76**, 122–134 (2016)
32. T. Sun, H. Pei, Y. Pan, C. Zhang, Robust adaptive neural network control for environmental boundary tracking by mobile robots. Int. J. Rob. Nonlin. Control **23**, 123–136 (2013)
33. E. Lughofer, *Evolving Fuzzy Systems-Methodologies, Advanced Concepts and Applications* (Springer, Berlin, 2011). ISBN: 978-3-642-18086-6
34. C.J.B. Macnab, Using RBFs in a CMAC to prevent parameter drift in adaptive control. Neurocomputing **205**, 45–52 (2016)
35. Q. Wu, X. Wang, Q. Shen, Research on dynamic modeling and simulation of axial-flow pumping system based on RBF neural network. Neurocomputing **186**, 200–206 (2016)
36. R. Yang, P.V. Er, Z. Wang, K.K. Tan, An FBF neural network approach towards precision motion system with selective sensor fusion. Neurocomputing **199**, 31–39 (2016)
37. C. Zhang, H. Wei, L. Xie, Y. Shen, K. Zhang, Direct interval forecasting of winds peed using radial basis function neural networks in a multi-objective optimization framework. Neurocomputing **205**, 53–63 (2016)
38. J.S. Roger-Jang, C.T. Sun, E. Mitzutani, *Neuro-Fuzzy and Soft Computing, a Computational Approach to Learning and Machine Intelligence* (Pearson College Div; first edition, 1997). ISBN: 0-13-261066-3
39. J.J. Rubio, A method with neural networks for the classification of fruits and vegetables. Soft Comput. **21**, 7207–7220 (2017)
40. J.J. Rubio, USNFIS: uniform stable neuro fuzzy inference system. Neurocomputing **262**, 57–66 (2017)
41. J.J. Rubio, Stability analysis for an on-line evolving neuro-fuzzy recurrent network, in *Evolving Intelligent Systems: Methodology and Applications* (John Willey and Sons, Hoboken; IEEE Press, Piscataway, 2010). Chapter 8, pp. 173–199. ISBN: 978-0-470-28719-4

Chapter 7
SOFMLS: Online Self-organizing Fuzzy Modified Least Square Network

1 Introduction

Both neural networks and fuzzy logic are universal estimators, which can approximate any nonlinear function to any prescribed accuracy, provided that sufficient hidden neurons or fuzzy rules are available. Recent results show that the fusion procedure of these two different technologies seems to be very effective for nonlinear system identification [1]. In the last few years, the application of fuzzy neural networks for nonlinear system identification has been a very active area [2–4]. Structure and parameters learning are involved in the identification of a system with fuzzy neural networks.

The system identification can be classified into two groups: (1) offline identification [5–10] and (2) online identification [11–21].

In offline identification, the update of the parameters and the structure take place only after the whole training dataset has been presented, i.e., only after each epoch. In this kind of identification, the structure learning is used to generate the fuzzy rules by trial-and-error approaches, like the unbiasedness criterion [8]. Several approaches generate fuzzy rules from numerical data. One of the most common methods for structure initialization is the uniform partitioning of each input variable into fuzzy sets, resulting in a fuzzy grid. This approach is followed in ANFIS [6]. In an earlier study [5], the Takagi-Sugeno model was used for designing several neuro fuzzy identifiers. This approach consists of two learning phases: (1) structure learning, which involves finding the most important subset of variables of all the possible ones, the partition of the input space, and determining the number of fuzzy rules, and (2) parameters learning, which involves approximating some unknown parameters by the parameter updating. The parameter updating is employed after the structure is decided. Most of the structure learning methods are based on data clustering, such as the fuzzy C-means clustering [22] and the mountain clustering [10].

J. de J. Rubio, *Stability Analysis of Neural Networks and Evolving Intelligent Systems*, https://doi.org/10.1007/978-3-031-87282-2_7

In online identification, structure and parameters learning are updated immediately after presentation of each input-output pair, i.e., after each iteration. The online identification also includes (1) structure learning and (2) parameters learning. For structure learning, the clustering methods are mainly used. In the clustering, to update fuzzy rules, distance from the centers of fuzzy rules, potentials of new data sample, and error from previous step are used. Different mechanisms are employed in constructing the structure. The resource allocating network (RAN) [18] uses a geometric growing criterion to update the fuzzy rules. The evolving fuzzy neural networks (EFuNNs) [15] use the difference between two membership vectors to update the fuzzy rules. The dynamic evolving neural fuzzy inference system (DENFIS) [16], the self-constructing neural fuzzy inference network (SONFIN) [13], and the recurrent self-organizing neural fuzzy inference network (RSONFIN) [14] use the distance to update the fuzzy rules. The evolving Takagi-Sugeno (ETS) model [11] uses the potential to update the fuzzy rules. The Takagi-Sugeno inference algorithm of an earlier study [23] considers input and output data to update the rules.

A self-constructing algorithm is no longer a practical system if the number of input-output pairs is large, because the number of rules grows even if some data are grouped into clusters. Therefore, a pruning method is needed. The self-constructing neural fuzzy networks mentioned earlier do not have a pruning method, even though they can be used for online learning. To extract fuzzy rules in a growing fashion from a large numerical database, some self-constructing fuzzy networks have been presented. It has been shown that the dynamic fuzzy neural network (DFNN) [19] approach provides good results, and the error reduction ratio of each radial basis function neuron is used to decide which radial basis function neurons are important to the network. Thus, the less important radial basis function neuron may be deleted. The general dynamic fuzzy neural network (GDFNN) proposed in [20] tries to give reasonable explanations for some predefined training parameters in DFNN. These methods, however, depend on the number of total training data. In an earlier study [11], it was considered that if a new datum, which is accepted as a focal point of a new rule, is too close to a previously existing rule, then the old rule is replaced by the new one. The self-organizing fuzzy neural network (SOFNN) [17] approach proposes a pruning method devised from the optimal brain surgeon (OBS) approach [24]. The basic idea of the SOFNN is to use the second derivative information to find the unimportant neuron. In the simplified method for learning in evolving Takagi-Sugeno fuzzy models (simpl_eTS) [12], the density as the population is considered, the population of each cluster is monitored, and if it amounts to less than 1% of the total data samples, that cluster is ignored. The cluster is ignored in the algorithm at this iteration, but the rule is not pruned; thus, the network cannot decrease. In the Sequential Adaptive Fuzzy Inference System (SAFIS) [25], one threshold parameter is used for adding a rule, and another threshold parameter is employed for pruning a rule as shown in this chapter; however, they do not use the concept of density.

On the other hand, the stability problem of fuzzy neural networks is important for online identification, and the neural fuzzy networks mentioned earlier do not guarantee the stability. It is well known that normal identification algorithms (e.g., gradient descent and least square) are stable in ideal conditions. However, in the presence of unmodeled dynamics, they may become unstable. The lack of robustness of the parameter identification was demonstrated earlier [26], and became a hot issue in the 1980s, when some robust modification techniques were suggested [27]. Some robust modifications must be applied to assure stability with respect to uncertainties. Projection operator is an effective tool to guarantee that the fuzzy identification is bounded [9, 27]. Input-to-state stability (ISS) approach is applied for nonlinear system identification, using the gradient descent algorithm for the fuzzy networks [21] and for the neural networks [28]. A double dead-zone is used to assure the stability of the identification error in the gradient descent algorithm [29]. On the other hand, the Lyapunov method is used to prove that a double dead-zone Kalman filter training is stable [30].

In this chapter, an online self-organizing fuzzy modified least square (SOFMLS) network is proposed to address these problems in the nonlinear system identification. Structure and parameter learning are active at the same time-step in the algorithm. The model is capable of perceiving the change in the actual system and adapting (self-organize) itself to the new situation. A new network that uses unidimensional membership functions for each rule is proposed, and it avoids the singularity produced by the widths in the antecedent part for online identification. It generates a new rule if the smallest distance between the new data and all the existing rules (the winner rule) is more than a prespecified radius, and it considers input and output data when a new rule is generated. To obtain faster parameter convergence, a modified least square algorithm is used in parameters learning to train the centers and the widths in the antecedent part and the centers in the consequent part. A new pruning algorithm based on the density is proposed, where the density indicates the number of elements for each rule. The rule that has the smallest density (the looser rule) in a selected number of iterations is pruned if the value of its density is smaller than a specified threshold. The stability of the proposed algorithm is proven, and the bound of the average of the identification error is found. The condition that led the algorithm to avoid the local minimum is found, and it is proven that the parameters error is bounded by the initial parameters error.

2 Network for Nonlinear Identification

Let us consider the following unknown discrete-time nonlinear system:

$$y(k-1) = f\left[X(k-1)\right], \tag{7.1}$$

where $X(k-1) = [x_1(k-1)\ldots x_N(k-1)] = [y(k-2),\ldots,y(k-n-1),u(k-2),\ldots,u(k-m-1)] \in \Re^N$ $(N = n+m)$ is the input vector, $|u(k-1)|^2 \leq \overline{u}$, $y(k-1)$ is the output of the plant, and f is an unknown nonlinear smooth function $f \in C^\infty$. A generic fuzzy model is presented as a collection of fuzzy rules in the following form (Mamdani fuzzy model [9]):

$$R_j : \text{IF } x_1 \text{ is } A_{1,j} \text{ and } x_2 \text{ is } A_{2,j} \text{ and } \ldots x_N \text{ is } A_{N,j} \\ \text{THEN } v \text{ is } B_j, \tag{7.2}$$

where $M(j = 1, 2 \ldots M)$ fuzzy IF THEN rules and N fuzzy sets are used for each rule to perform a mapping from an input linguistic vector $X(k-1) = [x_1(k-1) \ldots x_N(k-1)] \epsilon \Re^N$ $(N = n + m)$ to an output linguistic scalar $v \epsilon \Re$. $A_{1,j} \ldots A_{N,j}$, and B_j are the standard fuzzy sets. Each input variable x_i has N fuzzy sets. By using mean inference, center-average defuzzifier and center fuzzifier, called Sugeno fuzzy inference system with weighted average (FIS), the output of the fuzzy logic system can be expressed [7, 9] as

$$\begin{aligned} \widehat{y}(k-1) &= a(k-1)/b(k-1), \\ a(k-1) &= \sum_{j=1}^{M} v_j(k-1) z_j(k-1), \\ b(k-1) &= \sum_{j=1}^{M} z_j(k-1), \\ z_j(k-1) &= \exp\left[-\gamma_j^2(k-1)\right], \\ \gamma_j(k-1) &= \frac{\sum_{i=1}^{N} w_j(k-1)(x_i(k-1) - c_j(k-1))}{N}, \end{aligned} \tag{7.3}$$

where $x_i(k-1)$ are inputs of system (7.1), $(i = 1 \ldots N)$, $c_j(k-1)$ and $w_j(k-1) = \frac{1}{\sigma_j(k-1)}$ are the centers and the widths of the membership functions of the antecedent part, respectively, $j = 1 \ldots M$, $v_j(k-1)$ are the centers of the membership functions of the consequent part. Let us define the functions $\phi_j(k-1)$ from (7.3) as [28]

$$\phi_j(k-1) = z_j(k-1)/b(k-1). \tag{7.4}$$

Then (7.3) can be rewritten as follows (Fig. 7.1):

$$\widehat{y}(k-1) = \sum_{j=1}^{M} v_j(k-1)\phi_j(k-1) = V^T(k-1)\Phi(k-1), \tag{7.5}$$

where $V(k-1) = \left[v_j(k-1) \ldots v_M(k-1)\right]^T \epsilon \Re^M$ and $\Phi(k-1) = \left[\phi_j(k-1) \ldots \phi_M(k-1)\right]^T \epsilon \Re^M$.

Remark 7.1 The networks of many earlier studies [7, 9, 11, 17], and [25] use membership functions as shown in this study, but they use the following functions:

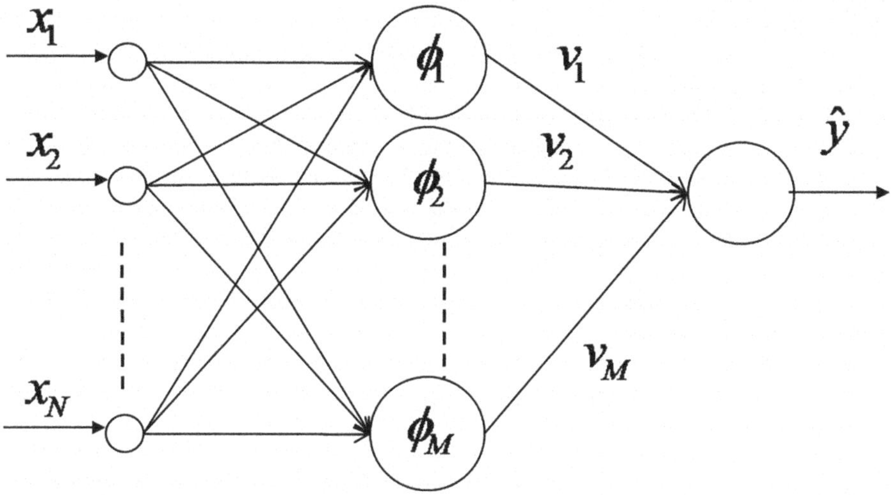

Fig. 7.1 Architecture of the fuzzy system

$$\gamma_j(k-1) = \sum_{i=1}^{N} \frac{1}{\sigma_{ij}(k-1)} \left(x_i(k-1) - c_{ij}(k-1)\right).$$

The first, in the antecedent part of the networks of the abovementioned references, $2N$ parameters are used for each membership function $(c_{ij}(k-1), \sigma_{ij}(k-1))$ called multidimensional membership functions, while in the antecedent part of the network in this study two parameters are used for each rule $(c_j(k-1), w_j(k-1))$, called unidimensional membership functions, as can be seen in (7.3). Second, the networks of the abovementioned references use $\frac{1}{\sigma_{ij}(k-1)}$ which can cause singularity in online learning, while the network of this study uses $w_j(k-1) = \frac{1}{\sigma_j(k-1)}$ to avoid singularity. Some authors use the sum inference [17], while some use the product inference [9, 11, 12, 16], and others employ the norm inference [7, 25]; however, in this study a new inference called mean inference is used. The mean inference is defined in (7.3) as $\gamma_j(k-1)$.

3 Structure Learning

Choosing an appropriate number of rules is important in the design of fuzzy neural systems, because too many rules result in a complex fuzzy neural system that may be unnecessary for the problem, whereas too few rules produce a less powerful fuzzy neural system, which may be insufficient to achieve the objective. The number of rules is seen as a design parameter. It is determined based on the input-output pairs and the number of elements of each rule. The basic idea is to group the input-output pairs into clusters and use one rule for one cluster, i.e., the number of rules equals the number of clusters.

One of the simplest clustering algorithms is the nearest neighborhood clustering algorithm. In this algorithm, the first datum is considered as the first center of the first cluster. If the distance between the new data and its nearest cluster is less than a prespecified value (the radius r), then the nearest cluster to the data is updated; otherwise, this datum is considered as a new cluster center. The details are given as follows.

Let $X(k-1)$ be the newly incoming pattern; then from (7.3) an auxiliary parameter $p(k-1)$ is obtained as

$$p(k-1) = \max_{1 \leq j \leq M} z_j(k-1). \tag{7.6}$$

If $p(k-1) \geq r$, where r is a prespecified radius, $r \in (0,1)$, then a rule is not generated. The winner rule j^* is presented in the algorithm when $z_j(k-1) = p(k-1)$. As the winner rule is a rule that increments its importance in the algorithm, its density must be increased and is updated as

$$d_{j^*}(k) = d_{j^*}(k) + 1. \tag{7.7}$$

If $p(k-1) < r$, then a new rule is generated and $M = M+1$. Once a new rule is generated, the next step is to assign initial centers and widths of the corresponding membership functions, and a new density with value of 1 is generated for this rule as follows:

$$\begin{array}{l} c_{M+1}(k) = \dfrac{\sum\limits_{i=1}^{N} x_i(k)}{N}, \quad w_{M+1}(k) = \dfrac{\sum\limits_{i=1}^{N} [x_i(k) - c_{j^*}(k)]}{N}, \\ v_{M+1}(k) = y(k) \quad d_{M+1}(k) = 1. \end{array} \tag{7.8}$$

The abovementioned algorithm will no longer be a practical system if the number of input-output pairs is large, because the number of rules (clusters) grows, even if some data are grouped into rules (clusters). Therefore, one needs a pruning method. A new pruning algorithm based on the density is proposed, where the density is the number of times that each rule is used in the algorithm. From (7.8), it can be seen that when a new rule is generated, its density starts with 1, and from (7.7) it can be seen that when a datum is grouped in an existing rule, the density of this rule is increased by 1. Thus, each cluster (rule) has its own density. The least important rule is the one that has the smallest density. After some iterations (ΔL), the least important rule is pruned if the value of its density is smaller than a prespecified threshold (d_u), i.e., this rule is unnecessary in the algorithm. The details are given as follows:

Each ΔL iteration, where $\Delta L \in N$, $d_{\min}(k)$ is considered as follows:

$$d_{\min}(k) = \min_{1 \le j \le M} d_j(k), \tag{7.9}$$

If $M \ge 2$ and $d_{\min}(k) \le d_u$, this rule is pruned, where $d_u \in N$ is the minimum selected density that is allowed. It is called the threshold parameter. Once a rule is pruned, the next step is to assign centers and widths of the corresponding membership functions. The looser rule j_* is presented in the algorithm when $d_j(k) = d_{\min}(k)$. The looser rule is the less important rule of the algorithm, if $j \le j_*$, nothing is modified, but if $j > j_*$, then all the parameters are moved to organize them as follows:

$$\begin{aligned} c_{j-1}(k) &= c_j(k), \quad & w_{j-1}(k) &= w_j(k), \\ v_{j-1}(k) &= v_j(k), \quad & d_{j-1}(k) &= d_j(k). \end{aligned} \tag{7.10}$$

In this way, the looser rule j_* is sent to the last element ($j = M$). For $j = M$, the looser rule is pruned as follows:

$$c_M(k) = 0, \quad w_M(k) = 0, \quad v_M(k) = 0, \quad d_M(k) = 0. \tag{7.11}$$

Consequently, M is updated as $M = M - 1$ to decrease the size of the network.

If $d_{\min}(k-1) > d_u$ or $M = 1$, then this rule is not pruned. If there is only one rule denoted as $M = 1$, then the algorithm cannot prune this rule.

Finally, L is updated as $L = L + \Delta L$.

Remark 7.2 The parameters L and ΔL are needed, because the pruning algorithm is not active at each iteration. The initial value of L is ΔL, and the pruning algorithm works at the first time when $k = L$, and consequently, L is increased by ΔL. The pruning algorithm works for each ΔL iteration. The parameter ΔL was determined empirically as $5d_u$; thus, the pruning algorithm only has d_u as the designing parameter.

Remark 7.3 It can be seen that the max of $z_j(k-1)$ is taken in (7.6). This idea was taken from the competitive learning of the ART recurrent neural network [7, 31] to obtain the winner rule (in the case of the ART network, it is the winner neuron).

Remark 7.4 In an earlier study [17], the second derivative of an objective function is used to find the unimportant rule. In this study, the density parameter is used to find the unimportant rule. In another study [12] the density as the population is considered, the population of each cluster is monitored, and if it amounts to less than 1% of the total data samples, the cluster is ignored at this iteration. The rule is ignored as $v_{d\min}(k) = 0$, and subsequently, this weight is ignored in the term $\hat{y}(k-1)$ of (7.5). The cluster is ignored in the algorithm at this iteration, but the rule is not pruned; thus, the network cannot decrease. In an earlier study [25], two threshold parameters are considered, one for adding rules and the other for removing rules; however, they did not use the density.

4 Parameters Learning

The stability of structure and parameters learning is needed, because this algorithm works online. First, the model is linearized, and later, the stability of the proposed algorithm is analyzed.

According to Stone-Weierstrass theorem [32], the unknown nonlinear function f of (7.1) is approximated as

$$y(k-1) = \sum_{j=1}^{M} v_j^* \phi_j^* (k-1) + \in_f = V^{*T} \Phi^* (k-1) + \in_f, \tag{7.12}$$

where $\in_f = y(k-1) - V^{*T} \Phi^*(k-1)$ is the modeling error, $\phi_j^*(k-1) = z_j^*(k-1)/b^*(k-1)$, $b^*(k-1) = \sum_{j=1}^{M} z_j^*(k-1)$, $z_j^*(k-1) = \exp\left[-\gamma_j^{*2}(k-1)\right]$, $\gamma_j^*(k-1) = \sum_{i=1}^{N} w_j^* \left(x_i(k-1) - c_j^*\right)$, where v_j^*, w_j^*, and c_j^* are the optimal parameters that can minimize the modeling error $\in_f$ [33].

First, the network model is linearized and will be used to define the parameters updating and to prove the stability of the proposed algorithm.

In the case of three independent variables, a smooth function has a Taylor series as

$$\begin{array}{l} f(\omega_1, \omega_2, \omega_3) = f(\omega_{10}, \omega_{20}, \omega_{30}) + \frac{\partial f(\omega_1, \omega_2, \omega_3)}{\partial \omega_1} \left(\omega_1 - \omega_{10}\right) + \\ \frac{\partial f(\omega_1, \omega_2, \omega_3)}{\partial \omega_2} \left(\omega_2 - \omega_{20}\right) + \frac{\partial f(\omega_1, \omega_2, \omega_3)}{\partial \omega_3} \left(\omega_3 - \omega_{30}\right) + R_f, \end{array} \tag{7.13}$$

where R_f is the remainder of the Taylor series. If we let ω_1, ω_2, and ω_3 correspond to $c_j(k-1)$, $w_j(k-1)$, and $v_j(k)$, ω_{10}, ω_{20}, ω_{30} correspond to c_j^*, w_j^*, and v_j^*, let us define $\widetilde{c}_j(k-1) = c_j(k-1) - c_j^*$, $\widetilde{w}_j(k-1) = w_j(k-1) - w_j^*$ and $\widetilde{v}_j(k-1) = v_j(k-1) - v_j^*$, and then the Taylor series is applied to linearize (7.3) and (7.5) as

$$\begin{array}{l} V^T(k-1)\Phi(k-1) = V^{*T}\Phi^*(k-1) + \sum_{j=1}^{M} \frac{\partial V^T(k-1)\Phi(k-1)}{\partial c_j(k-1)} \widetilde{c}_j(k-1) \\ + \sum_{j=1}^{M} \frac{\partial V^T(k-1)\Phi(k-1)}{\partial w_j(k-1)} \widetilde{w}_j(k-1) + \sum_{j=1}^{M} \frac{\partial V^T(k-1)\Phi(k-1)}{\partial v_j(k-1)} \widetilde{v}_j(k-1) + R_f. \end{array} \tag{7.14}$$

Considering (7.3), (7.4), and (7.5), and using the chain rule [9, 13, 29, 30], gives

$$\begin{aligned}
&\frac{\partial V^T(k-1)\Phi(k-1)}{\partial c_j(k-1)} \\
&= \frac{\partial V^T(k-1)\Phi(k-1)}{\partial a(k-1)} \frac{\partial a(k-1)}{\partial z_j(k-1)} \frac{\partial z_j(k-1)}{\partial \gamma_j(k-1)} \frac{\partial \gamma_j(k-1)}{\partial c_j(k-1)} \\
&+ \frac{\partial V^T(k-1)\Phi(k-1)}{\partial b(k-1)} \frac{\partial b(k-1)}{\partial z_j(k-1)} \frac{\partial z_j(k-1)}{\partial \gamma_j(k-1)} \frac{\partial \gamma_j(k-1)}{\partial c_j(k-1)} \\
&= 2\gamma_j(k-1)z_j(k-1)w_j(k-1)\frac{v_j(k-1)}{b(k-1)} \\
&-2\gamma_j(k-1)z_j(k-1)w_j(k-1)\frac{a(k-1)}{b^2(k-1)}, \\
&\frac{\partial V^T(k-1)\Phi(k-1)}{\partial c_j(k-1)} = 2\gamma_j(k-1)z_j(k-1)w_j(k-1)\frac{[v_j(k-1)-\widehat{y}(k-1)]}{b(k-1)},
\end{aligned}$$

$$\begin{aligned}
&\frac{\partial V^T(k-1)\Phi(k-1)}{\partial w_j(k-1)} \\
&= \frac{\partial V^T(k-1)\Phi(k-1)}{\partial a(k-1)} \frac{\partial a(k-1)}{\partial z_j(k-1)} \frac{\partial z_j(k-1)}{\partial \gamma_j(k-1)} \frac{\partial \gamma_j(k-1)}{\partial w_j(k-1)} \\
&+ \frac{\partial V^T(k-1)\Phi(k-1)}{\partial b(k-1)} \frac{\partial b(k-1)}{\partial z_j(k-1)} \frac{\partial z_j(k-1)}{\partial \gamma_j(k-1)} \frac{\partial \gamma_j(k-1)}{\partial w_j(k-1)} \\
&= -2\gamma_j(k-1)z_j(k-1)\frac{\sum_{i=1}^{N}[x_i(k-1)-c_j(k-1)]}{N}\frac{v_j(k-1)}{b(k-1)} \\
&+2\gamma_j(k-1)z_j(k-1)\frac{\sum_{i=1}^{N}[x_i(k-1)-c_j(k-1)]}{N}\frac{a(k-1)}{b^2(k-1)}, \\
&\frac{\partial V^T(k-1)\Phi(k-1)}{\partial w_j(k-1)} = 2\gamma_j(k-1)z_j(k-1)\frac{\sum_{i=1}^{N}[x_i(k-1)-c_j(k-1)]}{N}\frac{[\widehat{y}(k-1)-v_j(k-1)]}{b(k-1)},
\end{aligned}$$

$$\frac{\partial V^T(k-1)\Phi(k-1)}{\partial v_j(k-1)} = \frac{\partial \sum_{j=1}^{M} v_j(k-1)\phi_j(k-1)}{\partial v_j(k-1)} = \phi_j(k-1).$$

Substituting $\frac{\partial V^T(k-1)\Phi(k-1)}{\partial c_j(k-1)}$, $\frac{\partial V^T(k-1)\Phi(k-1)}{\partial w_j(k-1)}$, and $\frac{\partial V^T(k-1)\Phi(k-1)}{\partial v_j(k-1)}$ in (7.14) gives

$$\begin{aligned}
&V^T(k-1)\Phi(k-1) = V^{*T}\Phi^*(k-1) + \sum_{j=1}^{M}\phi_j(k-1)\widetilde{v}_j(k-1) + R_f \\
&+\sum_{j=1}^{M} 2\gamma_j(k-1)z_j(k-1)w_j(k-1)\frac{[v_j(k-1)-\widehat{y}(k-1)]}{b(k-1)}\widetilde{c}_j(k-1) \\
&+\sum_{j=1}^{M} 2\gamma_j(k-1)z_j(k-1)\frac{\sum_{i=1}^{N}[x_i(k-1)-c_j(k-1)]}{N}\frac{[\widehat{y}(k-1)-v_j(k-1)]}{b(k-1)}\widetilde{w}_j(k-1).
\end{aligned} \tag{7.15}$$

Let us define $B_j^c(k-1)$, $B_j^w(k-1)$, and $B_j^v(k-1)$ as

$$\begin{aligned} B_j^c(k-1) &= 2\gamma_j(k-1)z_j(k-1)w_j(k-1)\frac{[v_j(k-1)-\widehat{y}(k-1)]}{b(k-1)}, \\ B_j^w(k-1) &= -2\gamma_j(k-1)z_j(k-1)\frac{\sum_{i=1}^{N}[x_i(k-1)-c_j(k-1)]}{N}\frac{[\widehat{y}(k-1)-v_j(k-1)]}{b(k-1)}, \\ B_j^v(k-1) &= \phi_j(k-1). \end{aligned} \tag{7.16}$$

Using the abovementioned definitions in (7.15) gives

$$\begin{aligned} V^T(k-1)\Phi(k-1) &= V^{*T}\Phi^*(k-1) + \sum_{j=1}^{M} B_j^c(k-1)\widetilde{c}_j(k-1) \\ &+ \sum_{j=1}^{M} B_j^w(k-1)\widetilde{w}_j(k-1) + \sum_{j=1}^{M} B_j^v(k-1)\widetilde{v}_j(k-1) + R_f. \end{aligned} \tag{7.17}$$

Let us define the identification error as

$$e(k-1) = \widehat{y}(k-1) - y(k-1). \tag{7.18}$$

As $\sum_{j=1}^{M} B_j^c(k-1)\widetilde{c}_j(k-1)$, $\sum_{j=1}^{M} B_j^w(k-1)\widetilde{w}_j(k-1)$, and $\sum_{j=1}^{M} B_j^v(k-1)\widetilde{v}_j(k-1)$ are the product of two vectors, substituting (7.5), (7.12), and (7.18) in (7.17) gives

$$e\,(k-1) = B_{k-1}^T\widetilde{\theta}(k-1) + \mu(k-1), \tag{7.19}$$

where $B_{k-1}^T = \left[B_1^c(k-1), \ldots, B_M^c(k-1), B_1^w(k-1), \ldots, B_M^w(k-1), B_1^v(k-1), \ldots, B_M^v(k-1)\right] \in \Re^{1\times 3M}$, $\widetilde{\theta}(k-1) = [\widetilde{c}_1(k-1), \ldots, \widetilde{c}_M(k-1), \widetilde{w}_1(k-1), \ldots, \widetilde{w}_M(k-1), \widetilde{v}_1(k-1), \ldots, \widetilde{v}_M(k-1)]^T \in \Re^{3M\times 1}$, $\mu(k-1) = R_f - \epsilon_f$, $B_j^c(k-1)$, $B_j^w(k-1)$ and $B_j^v(k-1)$ are defined in (7.16). Thus, $\widetilde{\theta}_j(k-1) = \theta_j(k-1) - \theta_j^*$.

The least square [11, 16] is modified to get the stability of the algorithm, and subsequently, the modified least square to train the parameters and the structure is given as

$$\begin{aligned} \theta(k) &= \theta(k-1) - \frac{1}{Q_{k-1}}P_k B_{k-1} e(k-1), \\ P_k &= P_{k-1} - \frac{1}{R_{k-1}}P_{k-1}B_{k-1}B_{k-1}^T P_{k-1}, \end{aligned} \tag{7.20}$$

where $Q_{k-1} = R_2 + B_{k-1}^T P_{k-1} B_{k-1} \in \Re$, $R_{k-1} = 2Q_{k-1} + B_{k-1}^T P_{k-1} B_{k-1} \in \Re$, $0 < R_2 \in \Re$, B_{k-1} and $\theta(k-1)$ are given in (7.19), and it is assumed that the uncertainty is bounded [29, 30, 33], where $\overline{\mu}$ is the upper bound of the uncertainty

$\mu(k-1)$, $|\mu(k-1)| < \overline{\mu}$. $P_{k-1} \in \Re^{3M\times 3M}$ is a positive definite covariance matrix, $P_1 = cI$, where $c > 0$ is a scalar constant and $I \in \Re^{3M\times 3M}$ is the identity matrix.

The computational complexity of the algorithm is $O(N_c M^2)$ [34], where N_c is the size of R_{k-1}. The value of N_c is 1, while M is the size of $\theta(k)$ and is also the number of rules. The storage requirements are $O(M^2)$ [34]. It can be seen that the complexity requirements depend only on the number of rules. Therefore, it is important to have a low number of rules to have low memory requirements.

The following theorem gives the stability of the proposed algorithm.

Theorem 7.1 *The modified least square algorithm to train structure and parameters is uniformly stable, and the upper bound of the average error* $J(k-1)$ *satisfies*

$$\limsup_{T\to\infty} \frac{1}{T}\sum_{k=2}^{T} J(k-1) \leq \frac{\overline{\mu}^2}{R_2}, \tag{7.21}$$

where $J(k-1) = \frac{\left[B_{k-1}^T P_{k-1} B_{k-1}\right]^2}{Q_{k-1}^2 R_{k-1}} e^2(k-1)$.

Proof See [35] for the proof.

Remark 7.5 From (7.21) it can be seen that the final iteration parameter (time) T tends to infinity; thus, the stability of the proposed algorithm is preserved when $T \to \infty$.

Remark 7.6 The parameter M (number of rules) is finite, because the algorithm adds the necessary rules and prunes the unnecessary rules to adapt itself to the changing environment. The number of rules, M, is changed by the clustering and pruning algorithms, and M only changes the dimension of B_{k-1}^T and $\theta(k-1)$; thus, the stability result is preserved.

Remark 7.7 The theorem given in the study by Rubio and Yu [30] is very conservative, because it uses two dead-zones. However, Theorem 7.1 in this study is better, because it does not use any dead-zone.

Remark 7.8 The value of the parameter $\overline{\mu}$ is unimportant, because this parameter is not used in the algorithm. The bound of $\mu(k-1)$ is needed in order to guarantee the stability in the algorithm. This fact has been used in some earlier studies [29, 30, 33].

Corollary 7.1 *The parameters error* $\widetilde{\theta}(k)$ *is bounded as follows:*

$$\left\|\widetilde{\theta}(k)\right\|^2 \leq \left\|\widetilde{\theta}(1)\right\|^2, \tag{7.22}$$

where $\widetilde{\theta}(1)$ *is the initial parameters error.*

Proof See [35] for the proof.

Remark 7.9 From (7.22), it can be seen that for the modified least square algorithm, the parameters error is bounded by the initial parameters error. This result is better than that presented in earlier studies [21] and [28], because in these earlier studies they presented the stability analysis of a modified backpropagation algorithm for which even if the output error is convergent, it does not guarantee that the parameters error is bounded. Maybe, the parameters error can be very high and can make the system unstable.

Corollary 7.2 *The average error must satisfy* $J(k) \leq J(k-1)$ *to avoid the local minimum. The average error* $J(k-1)$ *of the modified least square algorithm avoids the local minimum when* $J(k-1) \geq \beta$, *where* $0 < \beta = \frac{\overline{\mu}^2}{R_2} + \frac{1}{c} \left\| \widetilde{\theta}(1) \right\|^2 < \infty$ *and* c *is defined in (7.20).*

Proof See [35] for the proof.

Remark 7.10 In an earlier study [36], a set of constraints to assure the interpretability of the membership functions has been given. These constraints help to avoid the local minimum, which is a problem of the backpropagation algorithm [7, 9, 13, 31, 36]. In this study, a modified least square algorithm is used [11, 12, 16, 17, 25, 30, 34]. The least square algorithm does not need to satisfy the constraints to assure the interpretability of the membership functions, because this algorithm does not have the problem of the local minimum, as can be seen in Corollary 7.2. In addition, the least square algorithm has faster parameters convergence than the backpropagation algorithm [11, 30, 34].

5 The Proposed Algorithm

The proposed algorithm is as follows:

1. Select the following parameters: the parameter of the modified least square algorithm is $R_2 > 0 \in \Re$, the parameter of the clustering algorithm is $0 < r < 1 \in \Re$, and the parameter of the pruning algorithm is $d_u \in N$, $(L = L + \Delta L$, $\Delta L = 5d_u)$.
2. For the first data $k = 1$ (where k is the number of iterations), $M = 1$ (where M is the number of rules or clusters), the initial parameters of the modified least square algorithm are $P_1 = cI \in \Re^{3M \times 3M}$ (where $0 < c \in \Re$), $v_1(1) = y(1)$, $c_1(1) = \frac{\sum_{i=1}^{N} x_i(1)}{N}$, and $w_1(1) = \text{rand} \in (0, 1)$ (v_1 is the initial parameter of the consequent part, c_1 and w_1 are the centers and widths of the membership function of the antecedent part), and the initial parameter of the clustering and pruning algorithms is $d_1(1) = 1$ (where d is the density parameter).
3. For the other data where $k \geq 2$, evaluate the fuzzy network parameters $z_j(k-1)$ and $b(k-1)$ with (7.3), evaluate the output of the fuzzy network $\widehat{y}(k-1)$ with (7.3), (7.4), and (7.5), evaluate the identification error $e(k-1)$ with (7.18), update the parameters of the modified least square algorithm $v_j(k)$, $c_j(k)$, and

$w_j(k)$ with (7.20), and evaluate the parameter of the clustering and pruning algorithm $p(k-1)$ with (7.6).

The updating of the clustering algorithm is as follows:

4. If $p(k-1) \geq r$, then a rule is not generated, the winner rule j^* is presented when $z_j(k-1) = p(k-1)$, and the value of the density $d_{j^*}(k)$ of this rule is updated with (7.7). The winner rule is a rule that increments its importance in the algorithm. Go to 3.
5. If $p(k-1) < r$, then a new rule is generated ($M = M+1$), where $r \in (0, 1)$ (e.g., the number of rules is increased by 1), the initial values of $c_{M+1}(k)$, $w_{M+1}(k)$, $v_{M+1}(k)$, and $d_{M+1}(k)$ are assigned to the new rule with (7.8), and the missing parameters are added to have $P_k \in \Re^{3(M+1)x3(M+1)}$ with diagonal elements (where P_k, $v_j(k)$, $c_j(k)$, and $w_j(k)$ are the parameters of the modified least square algorithm, and $d_j(k)$ is the parameter of the density, $j = 1 \ldots M$). Go to 3.

The updating of the pruning algorithm is as follows:

6. For the case where $k = L$, the pruning algorithm works (the pruning algorithm is not active at each iteration) and evaluates the minimum density $d_{\min}(k)$ with (7.9), and L is updated as $L = L + \Delta L$.
7. If $M \geq 2$ and $d_{\min}(k) \leq d_u$, then this rule is pruned, where $d_u \in N$ is the threshold of the density, and the looser rule j_* is presented when $d_j(k) = d_{\min}(k)$. The looser rule is the least important rule of the algorithm, the values of $c_j(k)$, $w_j(k)$, $v_j(k)$, and $d_j(k)$ are assigned with (7.10) and (7.11) to prune the looser rule j_*, and in the same way, the values of P_k are assigned to prune the looser rule j_* (where P_k, $v_j(k)$, $c_j(k)$, and $w_j(k)$ are the parameters of the modified least square algorithm and $d_j(k)$ is the parameter of the density, $j = 1 \ldots M$), and M is updated as $M = M - 1$ (e.g., the number of rules is decreased by 1). Go to 3.
8. If $d_{\min}(k) > d_u$ or $M = 1$, then this rule is not pruned. Go to 3.

6 Simulations

In this section, the suggested online self-organized algorithm is applied for nonlinear system identification. Note that in this study, the structure and parameters learning work at each time-step and they work online. The proposed network will be compared with networks that add and remove rules online, such as the Simpl_eTS [12], the SOFNN [17], and the SAFIS [25], because these networks have good performance.

Example 7.1 Let us consider the nonlinear system given and used in earlier studies [9, 25]:

$$y(k) = \frac{y(k-1)y(k-2)\left[y(k-1) - 0.5\right]}{1 + y^2(k-1) + y^2(k-2)} + u(k-1). \tag{7.23}$$

Table 7.1 Results for Example 7.1

Methods	No. of rules	Training RMSE	Testing RMSE
eTS($r = 1.8, \Omega = 10^6$)	49	0.0292	0.0212
Simpl_eTS ($r = 2.0, \Omega = 10^6$)	22	0.0528	0.0225
SAFIS ($\gamma = 0.997, \varepsilon_{\max} = 1, k = 1$) ($\varepsilon_{\min} = 0.1, e_g = 0.05, e_p = 0.005$)	17	0.0539	0.0221
SOFMLS	5	0.0341	0.0201

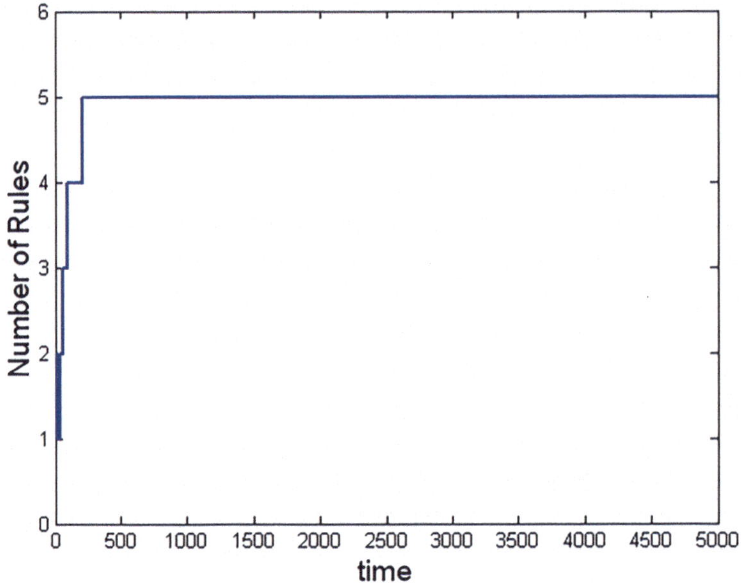

Fig. 7.2 Growth of rules for Example 7.1

As in the earlier studies [9, 25], the input $u(k)$ is given by $u(k) = \sin(2\pi k/25)$. The parameters of the SOFMLS are $P_1 = cI \in \Re^{3x3}$, where $c = 0.25$, $R_2 = 0.1$, $r = 0.9$, and $d_u = 6$. For the purpose of training and testing, 5000 and 200 data are produced, respectively. The average performance comparison of the SOFMLS with the eTS [11], the Simpl_eTS [12], and the SAFIS [25] is shown in Table 7.1, where the root mean square error (RMSE) [15] is

$$\text{RMSE} = \left(\frac{1}{N} \sum_{k=1}^{N} e^2(k-1) \right)^{\frac{1}{2}}. \tag{7.24}$$

From Table 7.1, it can be seen that the SOFMLS achieves better accuracy when compared with the other networks. In addition, the SOFMLS achieves this accuracy with the smallest number of rules. The evolution of the fuzzy rules for the SOFMLS for a typical run is shown in Fig. 7.2. From this figure, it can be seen that the SOFMLS produces five rules, and changes in the behavior are before 500 iterations.

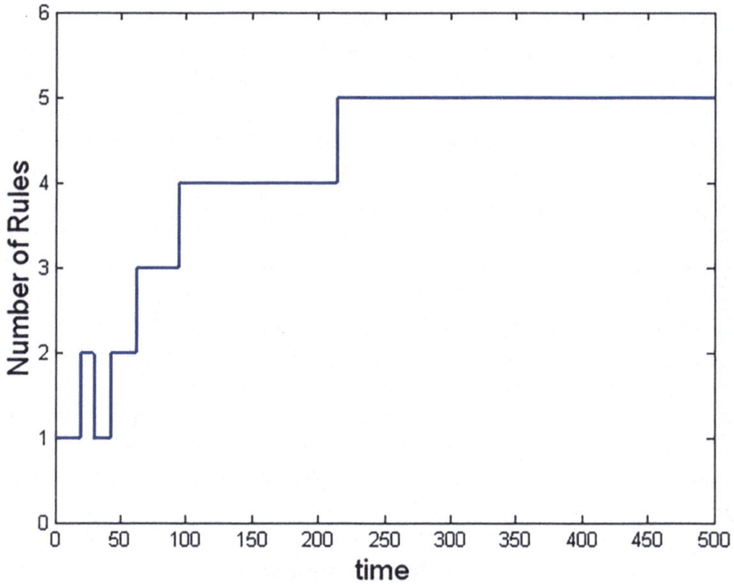

Fig. 7.3 Growth of rules for 500 iterations for Example 7.1

Figure 7.3 gives a clear illustration of the rule evolution tendency from 0 to 500 iterations and shows how the SOFMLS can automatically add and prune a rule during learning.

Figure 7.4 shows the resulting fuzzy membership functions in the antecedent part for the SOFMLS.

Figure 7.5 shows the evolution of the parameters c, w, and v for 1000 iterations for the SOFMLS. From this figure, it can be seen that some parameters appear when a new rule is added, and some disappear when a rule is pruned.

Figure 7.6 shows the average of the identification error for the SOFMLS. From this figure, it can be observed that the average of the identification error is bounded during training as in Theorem 7.1.

Figure 7.7 shows the testing result for the SOFMLS.

Example 7.2 Let us consider the nonlinear system given in an earlier study [9]:

$$y(k+1) = 0.3y(k) + 0.6y(k-1) + f(u(k)). \tag{7.25}$$

With $f(u(k)) = 0.6\sin(\pi u(k)) + 0.3\sin(3\pi u(k)) + 0.1\sin(5\pi u(k))$, the input is $u(k) = \sin(2\pi k/200)$. The initial parameters of the SOFMLS are $P_1 = cI \in \Re^{3x3}$, where $c = 0.35$, $R_2 = 0.1$, $r = 0.93$, and $d_u = 6$. For the purpose of training and testing, 3000 and 200 data are produced, respectively. The average performance comparison of the SOFMLS with the SAFIS [25] is shown in Table 7.2, where the RMSE of (7.24) is used.

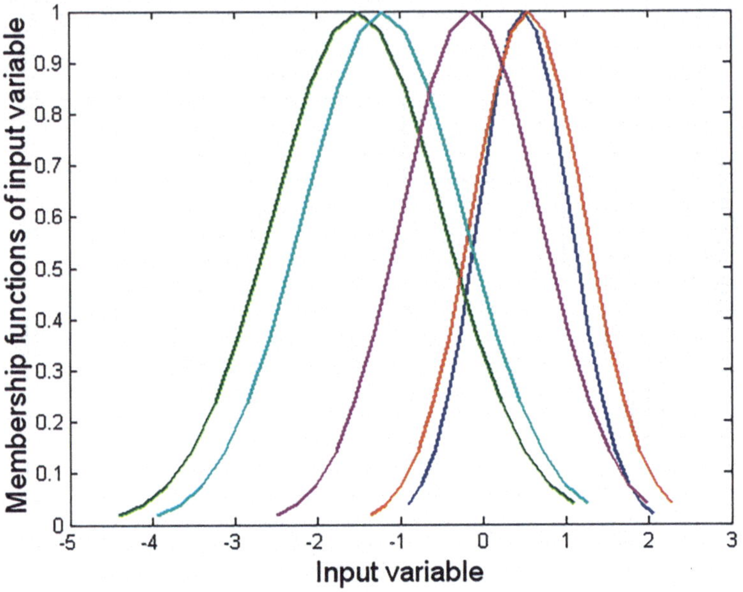

Fig. 7.4 Membership functions for Example 7.1

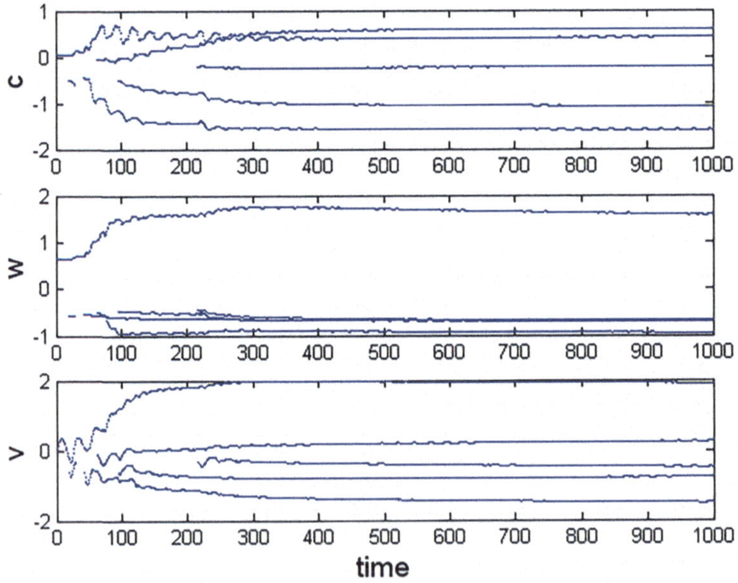

Fig. 7.5 Evolution of parameters of the network for Example 7.1

From Table 7.2, it can be seen that the SOFMLS achieves better accuracy when compared with the other network. In addition, the SOFMLS achieves this accuracy with the smallest number of rules. The evolution of the fuzzy rules for the SOFMLS

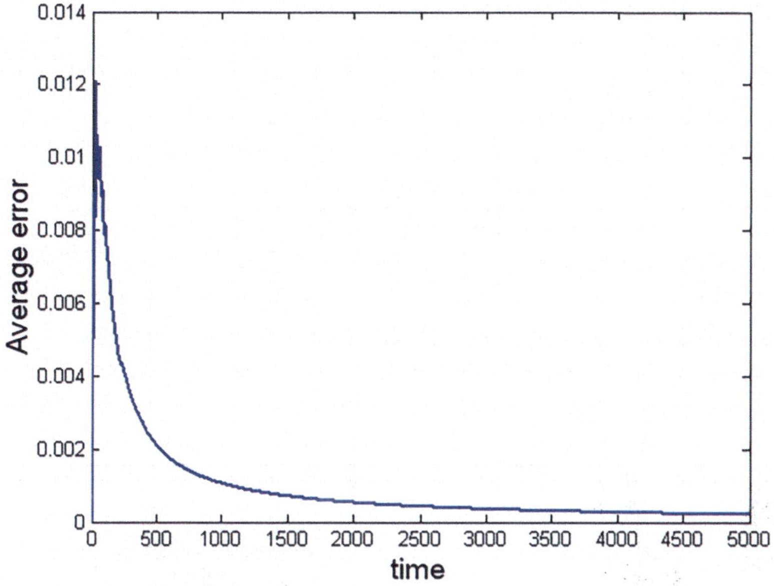

Fig. 7.6 Average error for Example 7.1

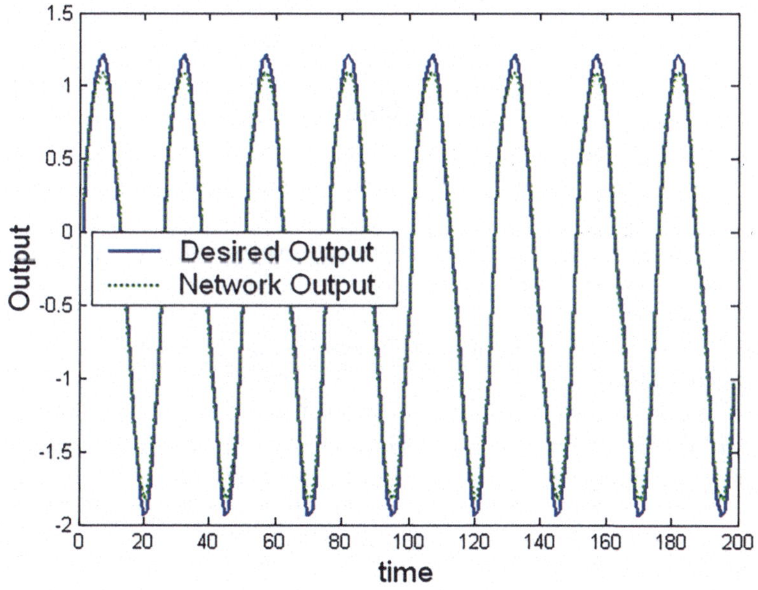

Fig. 7.7 Testing result for Example 7.1

and the SAFIS for a typical run are shown in Fig. 7.8. From this figure, it can be seen that the SOFMLS produces 6 rules and the SAFIS produces 11 rules.

Table 7.2 Results for Example 7.2

Methods	No. of rules	Training RMSE	Testing RMSE
SAFIS ($\gamma = 0.997$, $\varepsilon_{\max} = 2$, $k = 2$) ($\varepsilon_{\min} = 0.2$, $e_g = 0.03$, $e_p = 0.003$)	11	0.0507	0.0909
SOFMLS	6	0.0516	0.0290

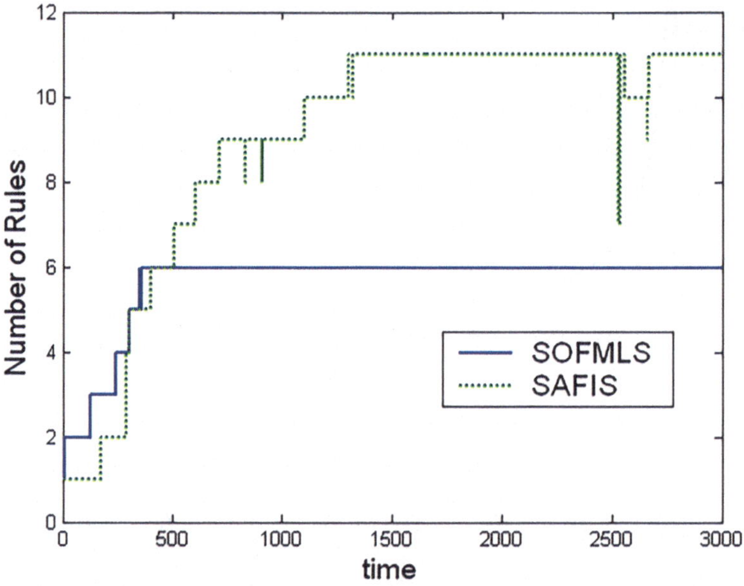

Fig. 7.8 Growth of rules for Example 7.2

Figure 7.9 gives a clear illustration for the rule evolution tendency from 0 to 500 iterations and also shows that both the networks can automatically add and prune a rule during learning.

Figure 7.10 shows the resulting fuzzy membership functions in the antecedent part for the SOFMLS.

Figure 7.11 shows the evolution of the parameters c, w, and v for 1000 iterations for the SOFMLS. From this figure, it can be seen that some parameters appear when a new rule is added and some disappear when a rule is pruned.

Figure 7.12 shows the average of the identification error for the SOFMLS. From this figure, it can be observed that the identification error is bounded during the training as in Theorem 7.1.

Figure 7.13 shows the testing result for the SOFMLS.

Example 7.3 The identification of the Box Jenkins furnace [37] is a well-known problem. There are originally 290 data pairs $(u(k), y(k))$. $y(k)$ is the output CO_2 concentration, and $u(k)$ is the input gas flow rate. A total of 200 samples are used for training, and the remaining 90 are used for testing. For the network, a series

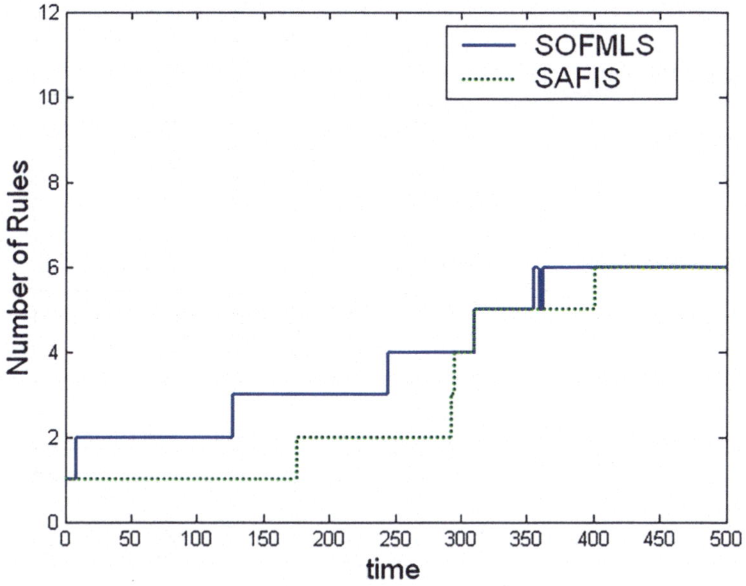

Fig. 7.9 Growth of rules for 500 iterations for Example 7.2

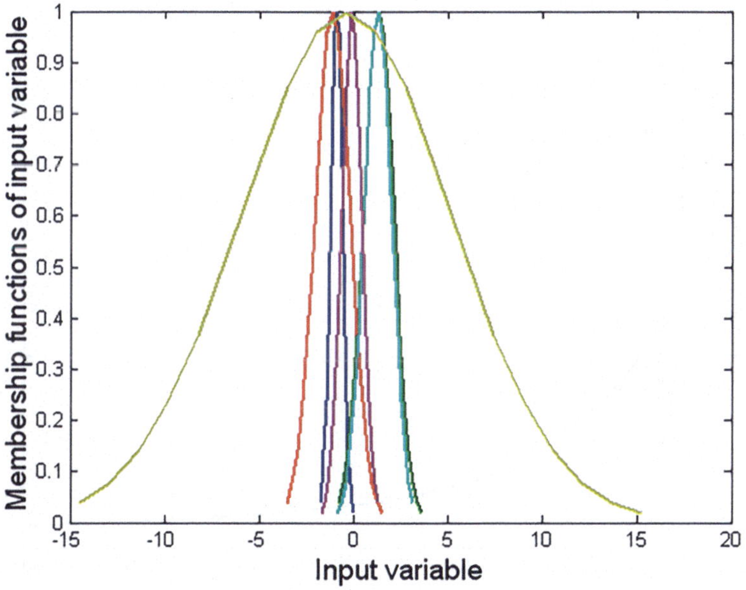

Fig. 7.10 Membership functions for Example 7.2

parallel model is used to model this system as $\widehat{y}(k) = f(y(k-1), u(k-4))$. The parameters of the SOFMLS are $P_1 = cI \in \Re^{3x3}$, where $c = 750$, $R_2 = 0.1$,

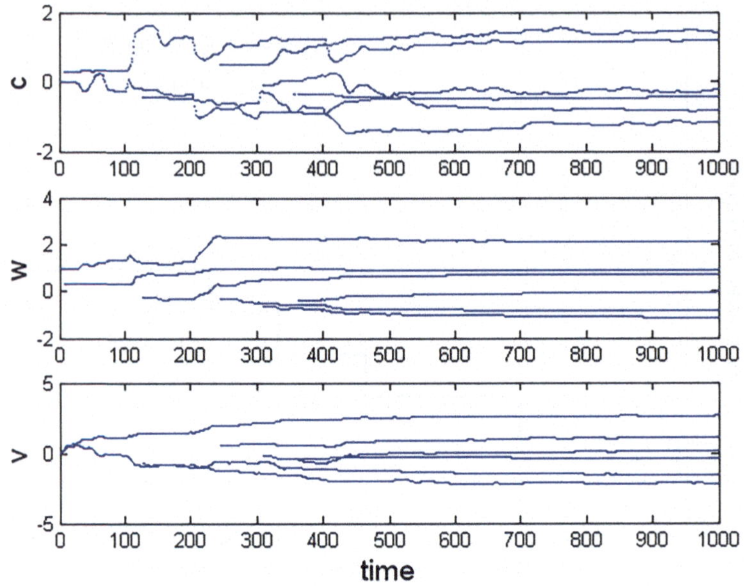

Fig. 7.11 Evolution of parameters of the network for Example 7.2

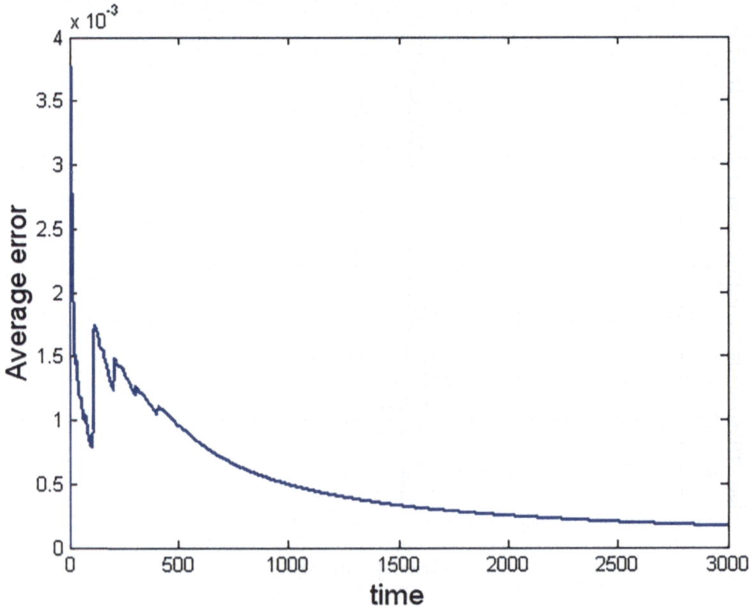

Fig. 7.12 Average error for Example 7.2

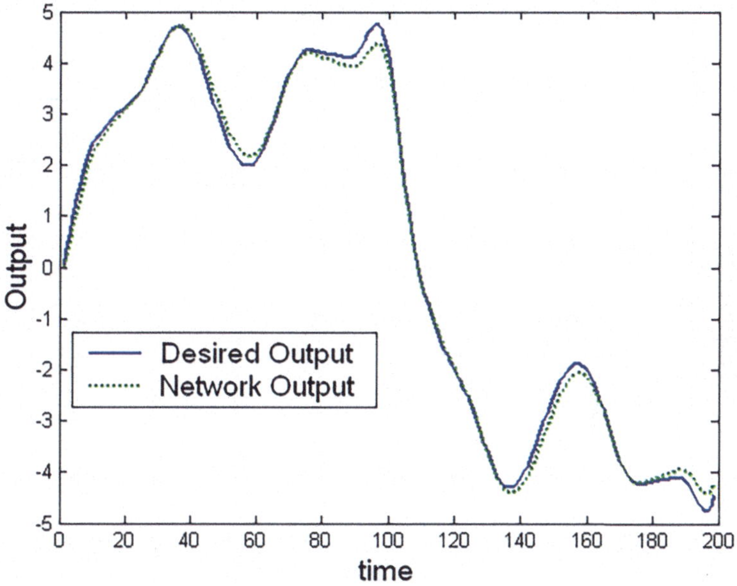

Fig. 7.13 Testing result for Example 7.2

Table 7.3 Results for Example 7.3

Methods	No. of rules	Testing RMSE
eTS($r = 0.4$, $\Omega = 750$)	5	0.0490
Simpl_eTS ($r = 0.4$, $\Omega = 750$)	3	0.0485
SOFNN ($\delta = .07$, $k_{rmse} = 0.3$, $k_d(1) = k_d(2) = 1.4$) ($\sigma_0 = 1$, $k_d(3) = k_d(3) = k_d(3) = k_d(3) = 0.5$)	4	0.0480
SOFMLS	5	0.0474

$r = 0.9$, and $d_u = 6$. The average performance comparison of the SOFMLS with the eTS [11], the Simpl_eTS [12], and the SOFNN [17] is shown in Table 6.3, where the RMSE of (7.24) is used.

From Table 7.3, it can be observed that the SOFMLS achieves better accuracy when compared with the other networks. In addition, the SOFMLS achieves this accuracy with a similar number of rules. The evolution of the fuzzy rules for the SOFMLS for a typical run is shown in Fig. 7.14. From this figure, it can be seen that the SOFMLS produces five rules.

Figure 7.15 shows the resulting fuzzy membership functions in the antecedent part for the SOFMLS.

Figure 7.16 shows the evolution of the parameters c, w, and v for the SOFMLS. From this figure, it can be seen that some parameters appear when a new rule is added and some disappear when a rule is pruned.

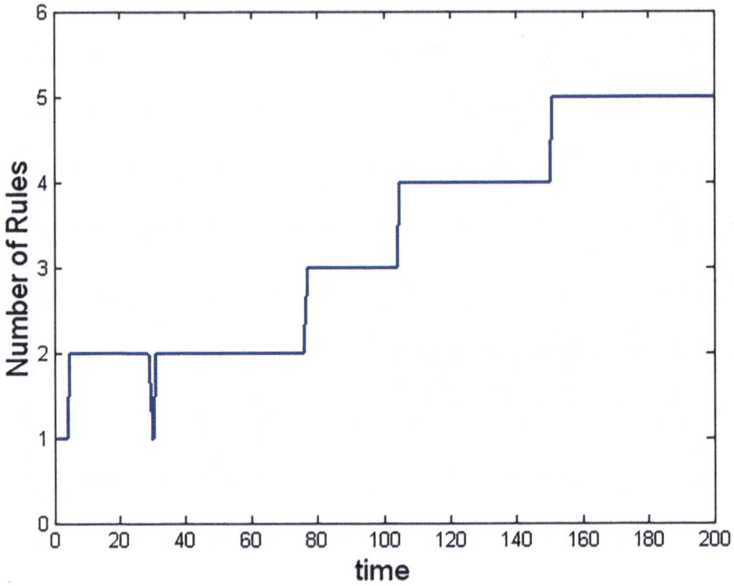

Fig. 7.14 Growth of rules for Example 7.3

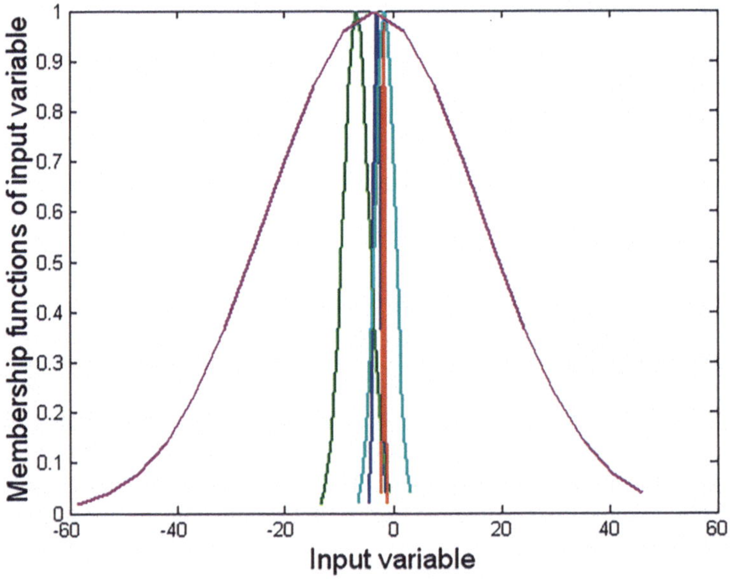

Fig. 7.15 Membership functions for Example 7.3

Figure 7.17 shows the average of the identification error for the SOFMLS. From this figure, it can be observed that the identification error during the training is bounded as in Theorem 7.1.

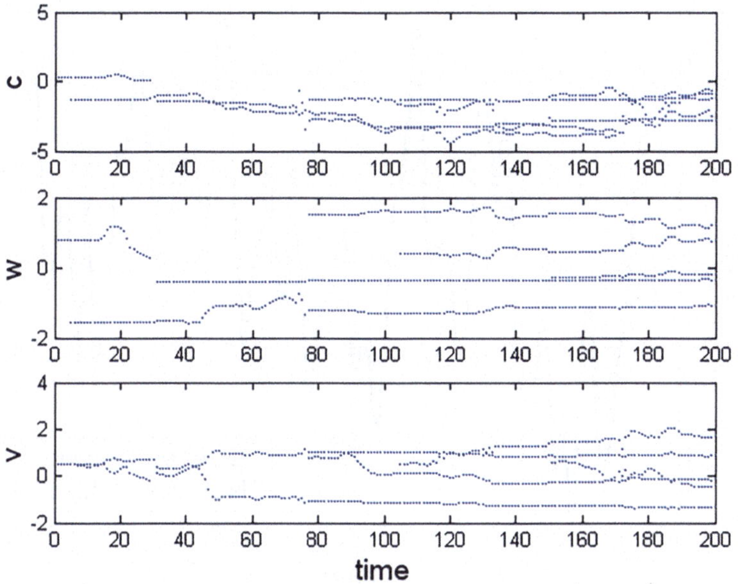

Fig. 7.16 Evolution of parameters of the network for Example 7.3

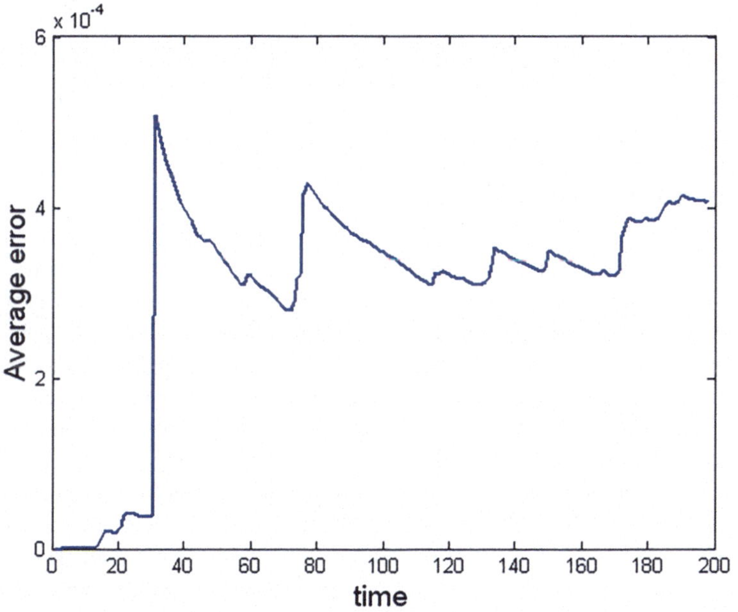

Fig. 7.17 Average error for Example 7.3

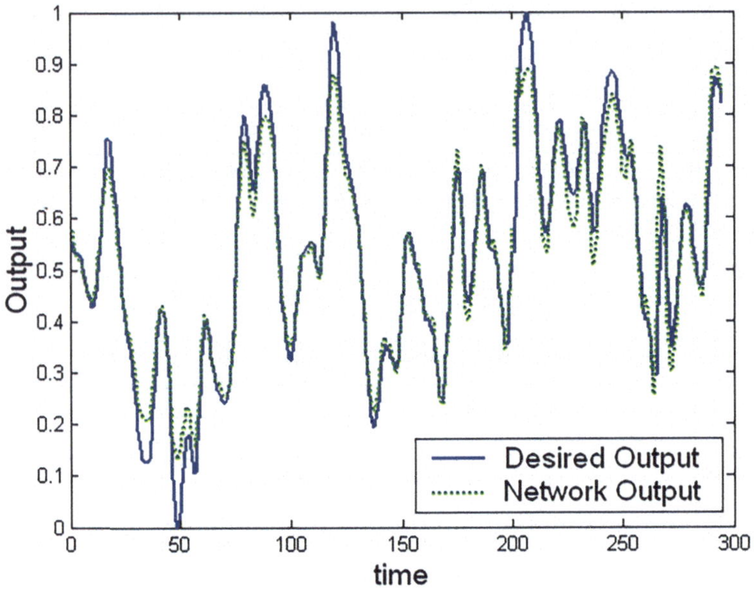

Fig. 7.18 Training and testing results for Example 7.3

Figure 7.18 shows the training and testing results for the SOFMLS; the first 200 samples belong of the training, and the remaining 90 are of the testing.

7 Concluding Remarks

In this chapter, a quick and efficient approach for system modeling using a fuzzy modified least square network is presented, which does not require retraining of the whole model. It is based on recursive building of the rule base by unsupervised and supervised learning, the rule-based model structure learning, and parameters estimation. The adaptive nature of this model, in addition to the transparent and compact form of fuzzy rules, makes it a promising candidate for online modeling and control of complex competitive processes with neural networks. From a dynamic system point of view, such training can be useful for all neural network applications requiring real-time updating of the weights. The main advantages of the approach are that (1) the network can develop an existing model when the data changes, (2) the network can start to learn a process from a single data sample and improve its performance through the time, and (3) it is recursive and highly effective. The proposed concept can be used in many fields, including nonlinear adaptive control, fault detection and diagnostics, performance analysis of dynamic systems, pattern and image recognition, time-series, identification of nonlinear systems, intelligent agents, and modeling. The results illustrate the viability, efficiency, and the potential of the approach when a limited amount of initial information is obtained. These characteristics are especially important in autonomous, robotics, and mechatronic systems.

References

1. M. Brown, C.J. Harris, *Adaptive Modelling and Control* (Macmillan Pub. Co., Prentice Hall, New York, 1994)
2. R. Babuka, H. Verbruggen, Neuro fuzzy methods for nonlinear system identification. Ann. Rev. Control **27**(1), 73–85 (2003)
3. W. Duch, R. Setiono, J.M. Zurada, Computational intelligence methods for rule data understanding. Proc. IEEE **92**(5), 771–805 (2004)
4. A. Sala, T.M. Gerra, R. Bakuska, Perspectives of fuzzy systems and control. Fuzzy Sets Syst. **156**(3), 432–444 (2005)
5. M.F. Azem, M. Hanmandlu, N. Ahmad, Structure identification of generalized adaptive neuro-fuzzy inference systems. IEEE Trans. Fuzzy Syst. **11**(6), 666–681 (2003)
6. J.S.R. Jang, ANFIS: adaptive-network-based fuzzy inference system. IEEE Trans. Syst. Man Cybern. **23**, 665–685 (1993)
7. J.S.R. Jang, C.T. Sun, *Neuro-Fuzzy and Soft Computing* (Prentice Hall, Hoboken, 1996)
8. I. Rivals, L. Personnaz, Neural network construction and selection in nonlinear modelling. IEEE Trans. Neural Networks **14**(4), 804–820 (2003)
9. L.X. Wang, *A Course in Fuzzy Systems and Control* (Prentice Hall, Englewood Cliffs, 1997)
10. R.R. Yager, D.P. Filev, Learning of fuzzy rules by mountain clustering, in *Proceedings of SPIE Conference on Application of Fuzzy Logic Technology*, Boston, pp. 246–254 (1993)
11. P.P. Angelov, D.P. Filev, An approach to online identification of Takagi-Sugeno fuzzy models. IEEE Trans. Syst. Man Cybern. **32**(1), 484–498 (2004)
12. P.P. Angelov, D.P. Filev, Simpl_ets: a simplified method for learning evolving Takagi-Sugeno fuzzy models, in *The International Conference on Fuzzy Systems*, pp. 1068–1072 (2005)
13. C.F. Juang, C.T. Lin, An online self-constructing neural fuzzy inference network and its applications. IEEE Trans. Fuzzy Syst. **6**(1), 12–32 (1998)
14. C.F. Juang, C.T. Lin, A recurrent self-organizing fuzzy inference network. IEEE Trans. Neural Networks **10**(4), 828–845 (1999)
15. N.K. Kasabov, Evolving fuzzy neural networks for supervised/unsupervised online knowledge-based learning. IEEE Trans. Syst. Man Cybern. **31**(6), 902–918 (2001)
16. N.K. Kasabov, Q. Song, DENFIS: dynamic evolving neural-fuzzy inference system and its application for time-series prediction. IEEE Trans. Fuzzy Syst. **10**(2), 144–154 (2002)
17. G. Leng, T.M. McGinnity, G. Prasad, An approach for online extraction of fuzzy rules using a self-organising fuzzy neural network. Fuzzy Sets Syst. **150**, 211–243 (2005)
18. J. Platt, A resource-allocating network for function interpolation. Neural Comput. **3**(2), 213–225 (1991)
19. S. Wu, M.J. Er, Dynamic fuzzy neural networks-a novel approach to function approximation. IEEE Trans. Syst. Man Cybern. Part B **30**, 358–364 (2000)
20. S. Wu, M.J. Er, A fast approach for automatic generation of fuzzy rules by generalized dynamic fuzzy neural networks. IEEE Trans. Fuzzy Syst. **9**, 578–594 (2001)
21. W. Yu, X. Li, Fuzzy identification using fuzzy neural networks with stable learning algorithms. IEEE Trans. Fuzzy Syst. **12**(3), 411–420 (2004)
22. J.C. Bezdek, Fuzzy Mathematics in Pattern Classification. PhD thesis, Applied Mathematics Center, Cornell University, Ithaca (1973)
23. K. Kim, E.J. Whang, C.-W. Park, E. Kim, M. Park, *A Tsk Fuzzy Inference Algorithm for Online Identification*, pp. 179–188 (Springer, Berlin, 2005)

24. D. Hassibi, D.G. Stork, Second order derivatives for network pruning. Adv. Neural Information Proces. Syst. **5**, 164–171 (1993)
25. H.J. Rong, N. Sundararajan, G.B. Huang, P. Saratchandran, Sequential adaptive fuzzy inference system (SAFIS) for nonlinear system identification and prediction. Fuzzy Sets Syst. **157**(9), 1260–1275 (2006)
26. B. Egardt, *Stability of Adaptive Controllers*. Lecture Notes in Control and Information Sciences, vol. 20 (Springer, Berlin, 1979)
27. P.A. Ioannou, J. Sun, *Robust Adaptive Control* (Prentice-Hall, Inc., Upper Saddle River, 1996)
28. W. Yu, X. Li, Discrete-time neuro-identification without robust identification. IEE Proc. Control Theory Appl. **150**(3), 311–316 (2003)
29. J.J. Rubio, W. Yu, A new discrete-time sliding-mode control with time-varying gain and neural identification. Int. J. Control **79**(4), 338–348 (2006)
30. J.J. Rubio, W. Yu, Nonlinear system identification with recurrent neural networks and dead-zone Kalman filter algorithm. Neurocomputing **70**, 2460–2466 (2007)
31. J.R. Hilera, V.J. Martines, *Redes Neuronales Artificiales, Fundamentos, Modelos y Aplicaciones* (Addison Wesley Iberoamericana, Boston, 1995)
32. G. Cybenco, Approximation by superposition of sigmoidal activation function. Math. Control Signals Syst. **2**, 303–314 (1989)
33. E.B. Kosmatopoulos, M.M. Polycarpou, M.A. Christodoulou, P.A. Ioannou, High-order neural network structures for identification of dynamic systems. IEEE Trans. Neural Networks **6**(2), 422–431 (1995)
34. G.V. Puskorius, L.A. Feldkamp, Neurocontrol of nonlinear dynamical systems with Kalman filter trained recurrent networks. IEEE Trans. Neural Networks **5**(2), 279–297 (1994)
35. J.J. Rubio, SOFMLS: online self-organizing fuzzy modified least square network. IEEE Trans. Fuzzy Syst. **17**(6), 1296–1309 (2009)
36. J. Valente de Oliveira, Semantic constrains for membership functions optimization. IEEE Trans. Syst. Man Cybern. Part A **29**(1), 128–138 (1999)
37. G.E.P. Box, G.M. Jenkins, *Time Series Analysis, Forecasting and Control* (Holden Day, San Francisco, 1976)

Chapter 8
Evolving Intelligent System for the Modeling of Nonlinear Systems with Dead-Zone Input

1 Introduction

Non-smooth nonlinear characteristics such as dead-zone, backlash, and hysteresis are common in actuators, sensors such as mechanical connections, hydraulic servo-valves, and electric servomotors; they also appear in biomedical systems. Dead-zone is one of the most important nonsmooth nonlinearities in many industrial processes, which can severely limit the system performance, and its study has been drawing much interest in the control community for a long time. Some important results are shown in [1–4], and [5]. In many works, controllers are proposed; however, a modeling system has not been introduced. The modeling system can be used for the failure prediction, disturbance rejection, trajectory generation, observer, and controller designs on the systems where the nonlinear behavior that includes the dead-zone is unknown.

On the other hand, the evolving intelligent systems are characterized by abilities to adjust their structure and parameters to the varying characteristics of the environment (with the term of environment embracing processes/phenomena in which the system has to interact or deal with the users using the system). Some important results are presented by [6–17], and [18]. From the above works, [6–11, 15–18] use interesting clustering algorithms, and [6–8, 11, 16, 17] present novel pruning algorithms as in this study; nevertheless, an evolving intelligent system for the modeling of recurrent nonlinear systems with dead-zone input is rarely presented.

J. de J. Rubio, *Stability Analysis of Neural Networks and Evolving Intelligent Systems*, https://doi.org/10.1007/978-3-031-87282-2_8

Finally, the stable intelligent systems are characterized to be systems where some kind of stability is guaranteed, i.e., if there is boundedness on inputs of the algorithm, then there is also boundedness on outputs. Some important studies are given by [3, 4, 18–27], and [28]. The aforementioned works do not consider the stability analysis of a recurrent evolving intelligent system for the modeling of a nonlinear system with dead-zone input.

In this chapter, a stable evolving intelligent system is addressed for the modeling of nonlinear systems with dead-zone input. In addition, the stability of the proposed algorithm is guaranteed.

The chapter is organized as follows. In Sect. 2, the nonlinear system with dead-zone input is presented. In Sect. 3, the evolving intelligent system is introduced. In Sect. 4, the evolving intelligent system is linearized. In Sect. 5, the structure updating of the evolving intelligent system is described. In Sect. 6, the stability of the above algorithm is guaranteed. In Sect. 7, the proposed algorithm is summarized. In Sect. 8, the proposed algorithm is used for the modeling of two synthetic problems. Section 9 presents conclusions and suggests future research directions.

2 Nonlinear System

In this study, the system which will be modeled is composed of a nonlinear plant preceded by an actuator with a nonsymmetric dead-zone in such a way that the dead-zone output is the input of the plant:

$$\begin{aligned} x_i(k) &= x_i(k-1) + T x_{i+1}(k-1), \quad i = 1, \ldots, n-1, \\ x_n(k) &= x_n(k-1) + T\left[f\left(x(k-1)\right) + g\left(x(k-1), u(k-1)\right)\right], \end{aligned} \tag{8.1}$$

where $i = 1 \ldots n$, $x_i(k)$ is the ith state, $x(k-1) = [x_1(k-1), x_2(k-1), \ldots, x_n(k-1)] \in \Re^n$, $u(k-1) \in \Re$ is the output of the dead-zone and input of the system, and $u(k-1)$ and $x(k-1)$ are known. f and g are the unknown nonlinear smooth functions. $T \in \Re$ is the sample time. The nonsymmetric dead-zone can be represented by

$$u(k-1) = DZ(v(k-1)) = \begin{cases} m_r\left(v(k-1) - b_r\right) & v(k-1) \geq b_r \\ 0 & b_l < v(k-1) < b_r \\ m_l\left(v(k-1) - b_l\right) & v(k-1) \leq b_l, \end{cases} \tag{8.2}$$

where m_r and m_l are the right and left constant slopes for the dead-zone characteristic, and b_r and b_l represent the right and left breakpoints. Note that $v(k-1) \in \Re$ is the input of the dead-zone.

The nonlinear system (8.1)–(8.2) can be rewritten in the multivariable Brunovsky form [3]:

$$\begin{aligned} x_i(k) &= x_i(k-1) + Tx_{i+1}(k-1), \quad i = 1, \ldots, n-1, \\ x_n(k) &= h_n\left[x(k-1), u(k-1)\right], \end{aligned} \tag{8.3}$$

where $i = 1 \ldots n$, $x_i\,(k)$ is the ith state, $u\,(k-1) \in \Re$ is the dead-zone output given by (8.2), $x\,(k-1) = [x_1\,(k-1)\,, x_2\,(k-1)\,, \ldots, x_n\,(k-1)] \in \Re^n$. $h_n \in \Re$ is an unknown nonlinear smooth function.

Remark 8.1 The nonlinear systems with dead-zone (8.3) are inspired by the actuators used to move the links of robotic systems which are second order systems with the Brunovsky form [29, 30].

3 Evolving Intelligent System

The following parallel [31, 32] recurrent neural network is used to model the nonlinear system (8.3):

$$\begin{aligned} \widehat{x}_i(k) &= \widehat{x}_i(k-1) + T\widehat{x}_{i+1}(k-1), \quad i = 1, \ldots, n-1, \\ \widehat{x}_n(k) &= s\widehat{x}_n(k-1) + \widehat{f}_{k-1} + \widehat{g}_{k-1}, \end{aligned} \tag{8.4}$$

where $i = 1 \ldots n$, $\widehat{f}_{k-1} = V_{1k-1}\sigma(k-1)$, $\widehat{g}_{k-1} = V_{2k-1}\phi(k-1)u(k-1)$, $\widehat{x}_i\,(k)$ represents the ith state of the neural network, $\widehat{x}\,(k) = [\widehat{x}_1\,(k)\,, \widehat{x}_2\,(k)\,, \ldots, \widehat{x}_n\,(k)] \in \Re^n$. The parameter $s \in \Re$ is a stable scalar (where its value should lie within the unit circle). The weights in the output layer are $V_{1k} \in \Re^{1\times m_1}$, $V_{2k} \in \Re^{1\times m_2}$. σ is m_1-dimensional vector function, and $\phi(\cdot) \in \Re^{m_2\times m_2}$ is a diagonal matrix, which are given as follows:

$$\begin{aligned} \sigma(k-1) &= \left[\sigma_1(k-1), \sigma_2(k-1), \cdots \sigma_{m_1}(k-1)\right]^T, \\ \phi(k-1) &= \operatorname{diag}\left[\phi_1(k-1), \phi_2(k-1), \cdots \phi_{m_2}(k-1)\right], \end{aligned} \tag{8.5}$$

where σ_i and ϕ_i are given later. Each input variable x_i has n fuzzy sets. From [33, 34], it is known, by using product inference, center-average defuzzifier and center fuzzifier, called Sugeno fuzzy inference system with weighted average (FIS), the output of the fuzzy logic system can be expressed as

$$\begin{aligned}
&\widehat{f}_{k-1} = \tfrac{a_1(k-1)}{b_1(k-1)}, \\
&a_1(k-1) = \sum_{j=1}^{m_1} v_{1j}(k-1) z_{1j}(k-1), \\
&b_1(k-1) = \sum_{j=1}^{m_1} z_{1j}(k-1), \\
&z_{1j}(k-1) = \exp\left[-\gamma_{1j}^2(k-1)\right], \\
&\gamma_{1j}(k-1) = w_{1j}(k-1)\left(\widehat{x}_n(k-1) - c_{1j}(k-1)\right), \\
&\widehat{g}_{k-1} = \tfrac{a_2(k-1)}{b_2(k-1)}, \\
&a_2(k-1) = \sum_{j=1}^{m_2} v_{2j}(k-1) z_{2j}(k-1) u_j(k-1), \\
&b_2(k-1) = \sum_{j=1}^{m_2} z_{2j}(k-1), \\
&z_{2j}(k-1) = \exp\left[-\gamma_{2j}^2(k-1)\right], \\
&\gamma_{2j}(k-1) = w_{2j}(k-1)\left(\widehat{x}_n(k-1) - c_{2j}(k-1)\right),
\end{aligned} \tag{8.6}$$

where $\widehat{x}_n(k-1)$ is the nth state of the system (8.4), $c_{1j}(k-1)$ and $w_{1j}(k-1)$ are the centers and the widths of the membership function of the antecedent part, respectively, $j = 1 \ldots m_1$, and $v_j(k-1)$ is the center of the membership function of the consequent part.

Remark 8.2 The weighted average radial basis function of [33] is again (8.6) where $\widehat{x}_n(k-1)$ is the state of the system (8.4), $c_{1j}(k-1)$ and $w_{1j}(k-1)$ are the centers and widths of the hidden layer, respectively, $j = 1 \ldots m_1$, and $v_j(k-1)$ are the weights of the output layer. In the radial basis function network of [33], $\frac{1}{\sigma_{1j}(k-1)}$ is used instead of $w_{1j}(k-1)$. In this study, $w_{1j}(k-1)$ is used instead of $\frac{1}{\sigma_{1j}(k-1)}$ to avoid singularity in the modeling process.

Define $\sigma_j(k-1)$ and $\phi_j(k-1)$ as follows:

$$\begin{aligned}
\sigma_j(k-1) &= z_{1j}(k-1)/b_1(k-1), \\
\phi_j(k-1) &= z_{2j}(k-1)/b_2(k-1).
\end{aligned} \tag{8.7}$$

The above functions are the same given in (8.5); therefore, (8.6) can be written as follows:

$$\begin{aligned}\widehat{f}_{k-1} &= \sum_{j=1}^{m_1} v_{1j}(k-1)\sigma_j(k-1) = V_{1,k-1}\sigma(k-1),\\ \widehat{g}_{k-1} &= \sum_{j=1}^{m_2} v_{2j}(k-1)\phi_j(k-1)u_j(k-1) = V_{2,k-1}\phi(k-1)u(k-1),\end{aligned} \tag{8.8}$$

where $V_{1,k-1} = \left[v_{11}(k-1)\ldots v_{1m_1}(k-1)\right]^T \epsilon \Re^{m_1}$ and $V_{2,k-1} = [v_{21}(k-1)\ldots v_{2m_2}(k-1)]^T \epsilon \Re^{m_2}$. The parameter m_1 is changing with the algorithm structure, while the parameter m_2 is fixed and it is the dimension of $u(k-1)$. See Fig. 8.1.

Remark 8.3 The proposed algorithm of Fig. 8.1 is different with the Kalman filter method of [25, 27, 35], and Fig. 8.2, for three reasons: The first reason is that the Kalman filter approximates all the functions, while the proposed algorithm approximates only the last function, i.e., the first $n-1$ states are linear and dependent of the n state because the system has the multivariable Brunovsky form [3]; therefore, only the last function gives the approximation of the system, and less computation is required. Other way to explain this fact is that the Kalman filter of [25, 27, 35] uses n algorithms, while the proposed technique uses only one algorithm to obtain the modeling of the system. The second reason is that in the Kalman filter method only the parameters are changing with the time, while in the proposed algorithm the parameters and structure are changing with the time. The third reason is that in the Kalman filter of [25, 27, 35], the series-parallel model is used [31, 32] where $x_n(k)$ of (8.3) is considered as the input of $\widehat{f}_{k-1}$ and $\widehat{g}_{k-1}$, while in this work, the parallel model is used [31, 32] where the state $\widehat{x}_n(k)$ of (8.4) is considered as the input of $\widehat{f}_{k-1}$ and $\widehat{g}_{k-1}$.

Remark 8.4 There are three differences between the proposed algorithm of Fig. 8.1 with the evolving method of [36] and Fig. 8.2. The first difference is that the evolving system of [36] approximates all the functions, while the proposed algorithm approximates only the last function, i.e., the first $n-1$ states are linear and dependent of the n state because the system has the multivariable Brunovsky form [3]; therefore, only the last function gives the approximation of the system, and less computation is required. Other way to explain this fact is that the evolving system of [36] uses n algorithms, while the proposed technique uses only one algorithm to obtain the modeling of the system. The second difference is that in the evolving method of [36], the series-parallel model is used [31, 32] where $x_n(k)$ of (8.3) is considered as the input of $\widehat{f}_{k-1}$ and $\widehat{g}_{k-1}$, while in this work, the parallel model is used [31, 32] where the state $\widehat{x}_n(k)$ of (8.4) is considered as the input of $\widehat{f}_{k-1}$ and $\widehat{g}_{k-1}$; consequently, less information of the system is required in the proposed algorithm. And finally, the third difference is that the evolving method of [36] is applied on biological nonlinear systems, while the proposed method is applied on nonlinear systems with dead-zone input.

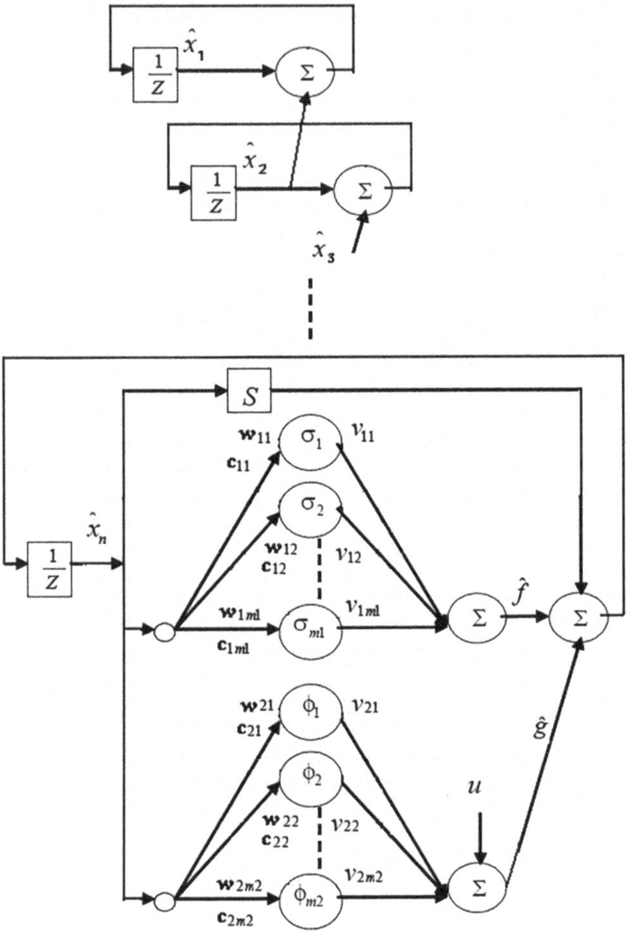

Fig. 8.1 Modified evolving intelligent system

4 Linearization of the Evolving Intelligent System

In this section, the model is linearized to find the parameters updating and to prove the stability of the proposed algorithm. The stability of the structure and output is required because this algorithm works on-line.

According to the Stone-Weierstrass theorem [37], the unknown nonlinear system (8.3) can be written in the following form:

$$
\begin{aligned}
x_i(k) &= x_i(k-1) + T x_{i+1}(k-1), \quad i = 1, \ldots, n-1, \\
x_n(k) &= s x_n(k-1) + f_{k-1} + g_{k-1},
\end{aligned}
\tag{8.9}
$$

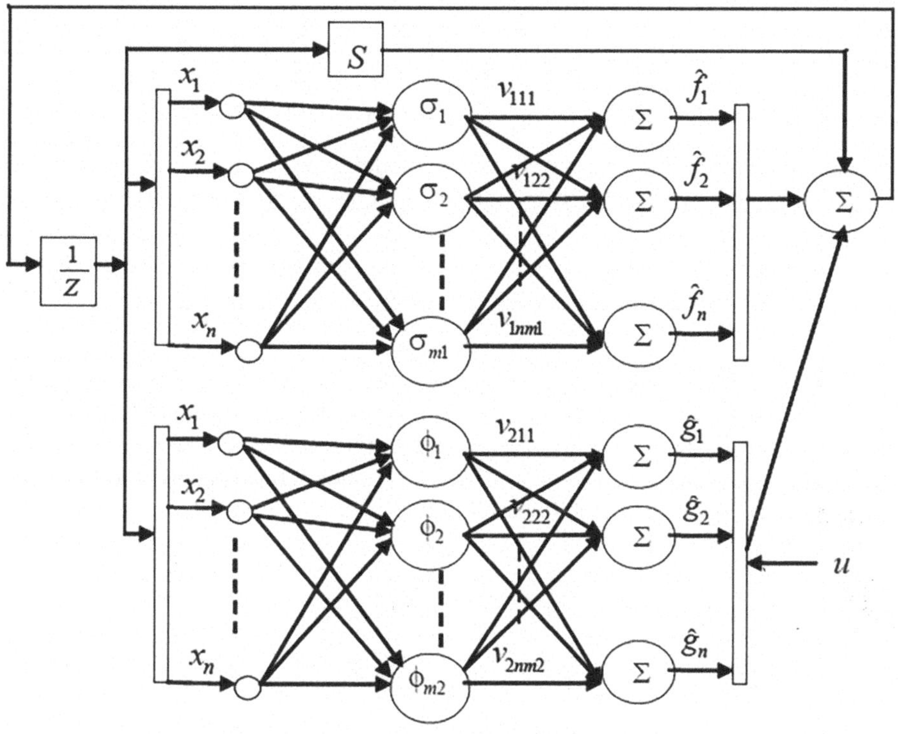

Fig. 8.2 Evolving intelligent system

where $f_{k-1} = V_{1,k-1}^{*}\sigma^{*}(k-1) + \epsilon_{k-1}^{f}$, $g_{k-1} = V_{2,k-1}^{*}\phi^{*}(k-1)u(k-1) + \epsilon_{k-1}^{g}$, $\epsilon_{k-1}^{f} + \epsilon_{k-1}^{g} = h_n\left[x(k), u(k)\right] - sx_n(k-1) - f_{k-1} - g_{k-1}$ represents unmodeled dynamics. By [37], it is known that the term $\epsilon_{k-1}^{f} + \epsilon_{k-1}^{g}$ can be made arbitrarily small by simply selecting an appropriate number of the hidden neurons. The unknown nonlinear function f_{k-1} of (8.9) is

$$f_{k-1} = \sum_{j=1}^{m_1} v_{1j}^{*}(k-1)\sigma_j^{*}(k-1) + \epsilon_{k-1}^{f} = V_{1,k-1}^{*}\sigma^{*}(k-1) + \epsilon_{k-1}^{f}, \tag{8.10}$$

where $\phi_j^*(k-1) = z_{1j}^*(k-1)/b_1^*(k-1)$, $b_1^*(k-1) = \sum_{j=1}^{m_1} z_{1j}^*(k-1)$, $z_{1j}^*(k-1) = \exp\left[-\gamma_{1j}^{*2}(k-1)\right]$, $\gamma_{1j}^*(k-1) = w_{1j}^*\left(x_n(k-1) - c_{1j}^*\right)$, v_{1j}^*, w_{1j}^*, and c_{1j}^* are the optimal parameters which can minimize the modeling error ϵ_{k-1}^f [37]. In the case of three independent variables, a smooth function has a Taylor series as follows:

$$\begin{array}{l} f(\alpha_1,\alpha_2,\alpha_3) = f(\alpha_{10},\alpha_{20},\alpha_{30}) + \frac{\partial f(\alpha_1,\alpha_2,\alpha_3)}{\partial \alpha_1}\left(\alpha_1 - \alpha_{10}\right) \\ +\frac{\partial f(\alpha_1,\alpha_2,\alpha_3)}{\partial \alpha_2}\left(\alpha_2 - \alpha_{20}\right) + \frac{\partial f(\alpha_1,\alpha_2,\alpha_3)}{\partial \alpha_3}\left(\alpha_3 - \alpha_{30}\right) + R_{k-1}^f, \end{array} \tag{8.11}$$

where R_{k-1}^f is the remainder of the Taylor series. Let α_1, α_2, and α_3 correspond to $c_{1j}(k-1)$, $w_{1j}(k-1)$, and $v_{1j}(k)$, α_{10}, α_{20}, and α_{30} correspond to c_{1j}^*, w_{1j}^*, and v_{1j}^*. Define $\widetilde{c}_{1j}(k-1) = c_{1j}(k-1) - c_{1j}^*$, $\widetilde{w}_{1j}(k-1) = w_{1j}(k-1) - w_{1j}^*$, and $\widetilde{v}_{1j}(k-1) = v_{1j}(k-1) - v_{1j}^*$. Thus, the Taylor series is applied to linearize $V_{1,k-1}\sigma(k-1)$ of (8.6) and (8.8) as follows:

$$\begin{array}{l} V_{1,k-1}\sigma(k-1) = V_{1,k-1}^*\sigma^*(k-1) + \sum_{j=1}^{m_1} \frac{\partial V_{1,k-1}\sigma(k-1)}{\partial c_{1j}(k-1)}\widetilde{c}_{1j}(k-1) \\ +\sum_{j=1}^{m_1} \frac{\partial V_{1,k-1}\sigma(k-1)}{\partial w_{1j}(k-1)}\widetilde{w}_{1j}(k-1) + \sum_{j=1}^{m_1} \frac{\partial V_{1,k-1}\sigma(k-1)}{\partial v_{1j}(k-1)}\widetilde{v}_{1j}(k-1) + R_{k-1}^f. \end{array} \tag{8.12}$$

Considering (8.6), (8.7), and (8.8) and using the chain rule [18, 22, 34], it gives

$$\begin{array}{l} \frac{\partial V_{1,k-1}\sigma(k-1)}{\partial c_{1j}(k-1)} = \frac{\partial V_{1,k-1}\sigma(k-1)}{\partial a_{1i}(k-1)} \frac{\partial a_{1i}(k-1)}{\partial z_{1j}(k-1)} \frac{\partial z_{1j}(k-1)}{\partial \gamma_{1j}(k-1)} \frac{\partial \gamma_{1j}(k-1)}{\partial c_{1j}(k-1)} \\ + \frac{\partial V_{1,k-1}\sigma(k-1)}{\partial b_1(k-1)} \frac{\partial b_1(k-1)}{\partial z_{1j}(k-1)} \frac{\partial z_{1j}(k-1)}{\partial \gamma_{1j}(k-1)} \frac{\partial \gamma_{1j}(k-1)}{\partial c_{1j}(k-1)} \\ = 2\gamma_{1j}(k-1)z_{1j}(k-1)w_{1j}(k-1)\frac{v_{1j}(k-1)-\widehat{f}_{k-1}}{b_1(k-1)}, \\ \frac{\partial V_{1,k-1}\sigma(k-1)}{\partial w_{1j}(k-1)} = \frac{\partial V_{1,k-1}\sigma(k-1)}{\partial a_{1i}(k-1)} \frac{\partial a_{1i}(k-1)}{\partial z_{1j}(k-1)} \frac{\partial z_{1j}(k-1)}{\partial \gamma_{1j}(k-1)} \frac{\partial \gamma_{1j}(k-1)}{\partial w_{1j}(k-1)} \\ + \frac{\partial V_{1,k-1}\sigma(k-1)}{\partial b_1(k-1)} \frac{\partial b_1(k-1)}{\partial z_{1j}(k-1)} \frac{\partial z_{1j}(k-1)}{\partial \gamma_{1j}(k-1)} \frac{\partial \gamma_{1j}(k-1)}{\partial w_{1j}(k-1)} \\ = 2\gamma_{1j}(k-1)z_{1j}(k-1)\left[\widehat{x}_n(k-1) - c_{1j}(k-1)\right]\frac{\widehat{f}_{k-1}-v_{1j}(k-1)}{b_1(k-1)}, \\ \frac{\partial V_{1,k-1}\sigma(k-1)}{\partial v_{1j}(k-1)} = \frac{\partial \sum_{j=1}^{m_1} v_{1j}(k-1)\sigma_j(k-1)}{\partial v_{1j}(k-1)} = \sigma_j(k-1). \end{array}$$

Substituting $\frac{\partial V_{1,k-1}\sigma(k-1)}{\partial c_{1j}(k-1)}$, $\frac{\partial V_{1,k-1}\sigma(k-1)}{\partial w_{1j}(k-1)}$, and $\frac{\partial V_{1,k-1}\sigma(k-1)}{\partial v_{1j}(k-1)}$ in Eq. (8.12), it gives

$$\begin{aligned} V_{1,k-1}\sigma(k-1) &= \sum_{j=1}^{m_1}\sigma_j(k-1)\widetilde{v}_{1j}(k-1) \\ &+\sum_{j=1}^{m_1}2\gamma_{1j}(k-1)z_{1j}(k-1)w_{1j}(k-1)\tfrac{v_{1j}(k-1)-\widehat{f}_{k-1}}{b_1(k-1)}\widetilde{c}_{1j}(k-1) \\ &+\sum_{j=1}^{m_1}2\gamma_{1j}(k-1)z_{1j}(k-1)\left[\widehat{x}_n(k-1)-c_{1j}(k-1)\right]\tfrac{\widehat{f}_{k-1}-v_{1j}(k-1)}{b_1(k-1)}\widetilde{w}_{1j}(k-1) \\ &+V^*_{1,k-1}\sigma^*(k-1)+R^f_{k-1}. \end{aligned} \tag{8.13}$$

Define $B^c_{1j}(k-1)$, $B^w_{1j}(k-1)$, and $B^v_{1j}(k-1)$ as

$$\begin{aligned} B^c_{1j}(k-1) &= 2\gamma_{1j}(k-1)z_{1j}(k-1)w_{1j}(k-1)\tfrac{v_{1j}(k-1)-\widehat{f}_{k-1}}{b_1(k-1)}, \\ B^w_{1j}(k-1) &= 2\gamma_{1j}(k-1)z_{1j}(k-1)\left[\widehat{x}_n(k-1)-c_{1j}(k-1)\right]\tfrac{\widehat{f}_{k-1}-v_{1j}(k-1)}{b_1(k-1)}, \\ B^v_{1nj}(k-1) &= \sigma_j(k-1). \end{aligned} \tag{8.14}$$

Note that $\sigma_j(k-1)$ is repeated for each i in $B^v_{1j}(k-1)$, and using the above definitions in (8.13), it gives

$$\begin{aligned} V_{1,k-1}\sigma(k-1) &= \sum_{j=1}^{m_1}B^c_{1j}(k-1)\widetilde{c}_{1j}(k-1) \\ &+\sum_{j=1}^{m_1}B^w_{1j}(k-1)\widetilde{w}_{1j}(k-1)+\sum_{j=1}^{m_1}B^v_{1j}(k-1)\widetilde{v}_{1j}(k-1) \\ &+V^*_{1,k-1}\sigma^*(k-1)+R^f_{k-1}. \end{aligned} \tag{8.15}$$

Define $\widetilde{f}_{k-1} = \widehat{f}_{k-1} - f_{k-1}$, and substituting (8.8), (8.10), and $\widetilde{f}_{k-1}$ into (8.15), it gives

$$\widetilde{f}_{k-1} = B^{fT}_{k-1}\widetilde{\theta}^f(k-1) + \mu^f(k-1), \tag{8.16}$$

where $B^{fT}_{k-1} = [B^c_{11}(k-1), \ldots, B^c_{1m_1}(k-1), B^w_{11}(k-1), \ldots, B^w_{1m_1}(k-1), B^v_{11}(k-1), \ldots, B^v_{1m_1}(k-1)] \in \Re^{1\times 3m_1}$, $\widetilde{\theta}^f(k-1) = [\widetilde{c}_{11}(k-1), \ldots, \widetilde{c}_{1m_1}(k-1), \widetilde{w}_{11}(k-1), \ldots, \widetilde{w}_{1m_1}(k-1), \widetilde{v}_{11}(k-1), \ldots, \widetilde{v}_{1m_1}(k-1)]^T \in \Re^{3m_1\times 1}$, thus $\widetilde{\theta}^f(k-1) = \theta^f(k-1) - \theta^{f*}$, $B^c_{1j}(k-1)$, $B^w_{1j}(k-1)$, and $B^v_{1j}(k-1)$ are given in (8.14), $\mu^f(k-1) = R^f_{k-1} - \epsilon^f_{k-1}$. Similarly in $\widetilde{g}_{k-1} = \widehat{g}_{k-1} - g_{k-1}$, it gives

$$\widetilde{g}_{k-1} = B^{gT}_{k-1}\widetilde{\theta}^g(k-1) + \mu^g(k-1), \tag{8.17}$$

where $B_{k-1}^{gT} = [B_{21}^{c}(k-1), \ldots, B_{2m_2}^{c}(k-1), B_{21}^{w}(k-1), \ldots, B_{2m_2}^{w}(k-1), B_{21}^{v}(k-1), \ldots, B_{2m_2}^{v}(k-1)] \in \Re^{1\times 3m_2}$, $\widetilde{\theta}^{g}(k-1) = [\widetilde{c}_{21}(k-1), \ldots, \widetilde{c}_{2m_2}(k-1), \widetilde{w}_{21}(k-1), \ldots, \widetilde{w}_{2m_2}(k-1), \widetilde{v}_{21}(k-1), \ldots, \widetilde{v}_{2m_2}(k-1)]^{T} \in \Re^{3m_2\times 1}$, thus $\widetilde{\theta}^{g}(k-1) = \theta^{g}(k-1) - \theta^{g*}$, $B_{2j}^{c}(k-1) = 2u_j(k-1)\gamma_{2j}(k-1)z_{2j}(k-1)w_{2j}(k-1)\frac{v_{2j}(k-1)-\widehat{g}_{k-1}}{b_2(k-1)}$, $B_{2j}^{w}(k-1) = 2u_j(k-1)\gamma_{2j}(k-1)z_{2j}(k-1)\left[\widehat{x}_n(k-1) - c_{2j}(k-1)\right]\frac{\widehat{g}_{k-1}-v_{2j}(k-1)}{b_2(k-1)}$, and $B_{2j}^{v}(k-1) = u_j(k-1)\phi_j(k-1)$, $\mu^{g}(k-1) = R_{k-1}^{g} - \epsilon_{k-1}^{g}$. Define the modeling error as follows:

$$e(k-1) = \widehat{y}(k-1) - y(k-1), \tag{8.18}$$

where $\widehat{y}(k-1) = \widehat{f}_{k-1} + \widehat{g}_{k-1} = \widehat{x}_n(k) - s\widehat{x}_n(k-1)$ is the network output and $y(k-1) = f_{k-1} + g_{k-1} = x_n(k) - sx_n(k-1)$ is the nonlinear system output, $e(k-1) \in \Re$; therefore, substituting $\widehat{y}(k-1)$, $y(k-1)$, (8.16), and (8.17) in (8.18), it gives

$$e\,(k-1) = B_{k-1}^{T}\widetilde{\theta}(k-1) + \mu(k-1), \tag{8.19}$$

where $B_{k-1}^{T} = [B^{gT}(k-1), B^{fT}(k-1)] \in \Re^{1\times 3(m_2+m_1)}$, $\widetilde{\theta}(k-1) = [\widetilde{\theta}^{g}(k-1), \widetilde{\theta}^{f}(k-1)]^{T} \in \Re^{3(m_2+m_1)\times 1}$, $\mu(k-1) = \mu^{g}(k-1) + \mu^{f}(k-1) \in \Re$, $\widetilde{\theta}^{f}(k-1)$ and $B^{fT}(k-1)$ are given in (8.16), and $\widetilde{\theta}^{g}(k-1)$ and $B^{gT}(k-1)$ are given in (8.17).

Define the state error as $\widetilde{x}_n(k) = \widehat{x}_n(k) - x_n(k)$. From (8.18), it gives $\widehat{x}_n(k) = s\widehat{x}_n(k-1) + \widehat{y}(k-1)$ and $x_n(k) = sx_n(k-1) + y(k-1)$, and subtracting the second equation to the first gives $\widehat{x}_n(k) - x_n(k) = s\left[\widehat{x}_n(k-1) - x_n(k-1)\right] + [\widehat{y}(k-1) - y(k-1)]$. Substituting $\widetilde{x}_n(k)$ and $e(k-1)$ of (8.18) in the above equation gives

$$\widetilde{x}_n(k) = s\widetilde{x}_n(k-1) + e(k-1). \tag{8.20}$$

5 Structure Updating

Choosing an appropriate number of hidden neurons is important in designing evolving intelligent systems, because too many hidden neurons result in a complex evolving system that may be unnecessary for the problem, and it can cause overfitting [33], whereas too few hidden neurons produce a less powerful neural system that may be insufficient to achieve the objective. The number of hidden neurons is considered as a design parameter, and it is determined based on the input-output pairs and on the number of elements of each hidden neuron. The basic idea is to group the input-output pairs into clusters and use one hidden neuron for one cluster; i.e., the number of hidden neurons equals the number of clusters [6–11, 15–18, 22].

One of the simplest clustering algorithms is the nearest neighborhood clustering algorithm. In this algorithm, the first data are considered as the center of the first cluster. Then, if the distances from a datum to the cluster centers are less than a prespecified value (the radius r), this datum is set into the closest cluster; otherwise, set this datum as a new cluster center. The details are given as follows.

Consider $x_n(k-1)$ as a newly incoming pattern; then from (8.6) it is obtained:

$$p(k-1) = \max_{1 \leq j \leq m1} z_{1j}(k-1). \tag{8.21}$$

If $p(k-1) < r$, then a new hidden neuron is generated (each hidden neuron corresponds to each center), and $m_1 = m_1 + 1$, where r is a selected radius, $r \in (0, 1)$. Once a new hidden neuron is generated, the next step is to assign initial centers and widths of the network, and a new density with value 1 is generated for this hidden neuron.

$$\begin{aligned} c_{1,m_1+1}(k) &= x_n(k), \quad w_{1,m_1+1}(k) = \text{rand}, \\ v_{1m_1+1}(k) &= y(k), \quad d_{m_1+1}(k) = 1. \end{aligned} \tag{8.22}$$

If $p(k-1) \geq r$, then a hidden neuron is not generated. If $z_{1j}(k-1) = p(k-1)$, the winner neuron j^* is obtained, and the winner neuron is a neuron that increments its importance in the algorithm, then its density must be increased and is updated as follows:

$$d_{j^*}(k) = d_{j^*}(k) + 1. \tag{8.23}$$

The above algorithm is no longer a practical system if the number of input-output pairs is large because the number of hidden neurons (clusters) grows, even some data are grouped into hidden neurons (clusters). Therefore, a pruning method is required [6–8, 11, 16, 17, 22]. The pruning algorithm is based on the density where the density is the number of times each hidden neuron is used in the algorithm. From (8.22), it is obtained that when a new hidden neuron is generated, its density starts at one, and from (8.23), it is known that when a datum is grouped in an existing hidden neuron, the density of this hidden neuron is increased by one. Then, each cluster (hidden neuron) has its own density. The least important hidden neuron is the hidden neuron which has the smallest density. After some iterations (ΔL) the least important hidden neuron is pruned if the value of its density is smaller than a specified umbral (d_u). The details are given as follows.

Each ΔL iterations where $\Delta L \in \aleph$, consider

$$d_{\min}(k) = \min_{1 \leq j \leq m_1} d_j(k), \tag{8.24}$$

If $m_1 \geq 2$ (if there is one hidden neuron given as $m_1 = 1$, the hidden neuron cannot be pruned) and if $d_{\min}(k) \leq d_u$, this hidden neuron is pruned, where $d_u \in \aleph$ is the minimum selected allowed density, and it is called the umbral parameter. Once a hidden neuron is pruned, the next step is to assign centers and widths of the network. When $d_j(k) = d_{\min}(k)$ the looser neuron j_* is obtained, the looser neuron is the least important neuron of the algorithm, if $j \leq j_*$ do nothing, but if $j > j_*$ all the parameters are updated as follows:

$$\begin{aligned} c_{1,j-1}(k) &= c_{1,j}(k), \quad w_{1,j-1}(k) = w_{1,j}(k), \\ v_{1j-1}(k) &= v_{1j}(k), \quad d_{j-1}(k) = d_j(k). \end{aligned} \tag{8.25}$$

The above parameters updating moves the looser neuron j_* to the last element ($j = m_1$). For $j = m_1$, the looser neuron is pruned as follows:

$$c_{1,m_1}(k) = 0, \quad w_{1,m_1}(k) = 0, \quad v_{1m_1}(k) = 0, \quad d_{m_1}(k) = 0. \tag{8.26}$$

Then m_1 is updated as $m_1 = m_1 - 1$ to decrease the size of the network.

If $d_{\min}(k-1) > d_u$ or $m_1 = 1$, do nothing.

Finally L is updated as $L = L + \Delta L$.

Remark 8.5 The parameters L and ΔL are because the pruning algorithm does not work in each iteration. The initial value of L is ΔL, the pruning algorithm works at the first time when $k = L$, and then L is increased by ΔL. The pruning algorithm works each ΔL iteration. The parameter ΔL is found empirically as $5d_u$; thus, the pruning algorithm only has d_u as the design parameter.

6 Stability Analysis

First, an important definition and theorem are mentioned. Later, the main stability theorem is presented.

Consider the following discrete-time nonlinear system:

$$x_{k+1} = f\left[x_k, u_k\right], \tag{8.27}$$

where $u_k \in \Re^m$ is the input vector, $x_k \in \Re^n$ is the state vector, and u_k and x_k are known. f is an unknown nonlinear smooth function $f \in C^\infty$.

Definition 8.1 The system (8.27) is said to be uniformly stable if $\forall \epsilon > 0, \exists \delta = \delta(\epsilon)$ such that

$$\|x_{k1}\| < \delta \Rightarrow \|x_k\| < \epsilon, \quad \forall k > k_1. \tag{8.28}$$

If the system has $\delta = \delta(\epsilon, k)$, then the system (8.27) is simply stable.

Now, a theorem for the stability of discrete-time nonlinear systems taken from [22] will be given.

Theorem 8.1 *Let $L_k(x(k))$ be a Lyapunov-like function of the discrete-time nonlinear system (8.27), if it satisfies*

$$\begin{aligned} \gamma_1(\|x_k\|) \leq L_k(x_k) \leq \gamma_2(\|x_k\|), \\ \Delta L_k(x_k) \leq -\gamma_3(\|x_k\|) + \gamma_3(\delta), \end{aligned} \tag{8.29}$$

where δ is a positive constant, $\gamma_1(\cdot)$ and $\gamma_2(\cdot)$ are K_∞ functions, and $\gamma_3(\cdot)$ is a K function; then the system (8.27) is uniformly stable.

Proof *See [22] for the proof.*

Remark 8.6 The continuous-time version of the above theorem is given by [38]. The main difference between the continuous-time stability theorem of [38] and the discrete-time stability theorem of [22] is that in the first, the derivative of the Lyapunov function is used, and in the second, the difference of the Lyapunov-like function is used.

Now, the stability of the proposed algorithm is analyzed.

Theorem 8.2 *Consider the evolving intelligent system (8.4), (8.6), (8.22), (8.30) to model the nonlinear systems with dead-zone input (8.1), (8.2), (8.3), and use the recursive least square updating function:*

$$\begin{aligned} \theta(k) &= \theta(k-1) - \frac{1}{Q_{k-1}} P_k B_{k-1} e(k-1), \\ P_k &= P_{k-1} - \frac{1}{R_{k-1}} P_{k-1} B_{k-1} B_{k-1}^T P_{k-1}, \end{aligned} \tag{8.30}$$

where $Q_{k-1} = 10 + B_{k-1}^T P_{k-1} B_{k-1}$, $R_{k-1} = 2Q_{k-1} + B_{k-1}^T P_{k-1} B_{k-1}$, B_{k-1}^T and $\theta(k-1)$ are given in (8.19), and $P_{k-1} \in \Re^{3(m_2+m_1)\times 3(m_2+m_1)}$ is a positive definite covariance matrix. Therefore, the average error of the modeling error is uniformly stable and will converge to

$$\limsup_{T\to\infty} \sum_{k=2}^{T} \frac{\left(B_{k-1}^T P_{k-1} B_{k-1}\right)^2}{Q_{k-1}^2 R_{k-1}} e^2(k-1) \leq \frac{\overline{\mu}}{10}, \tag{8.31}$$

where $\overline{\mu}$ is the upper bound of the uncertainty $\mu(k-1)$, $|\mu(k-1)| < \overline{\mu}$.

Proof See [39] for the proof.

Remark 8.7 The parameter m_1 (number of neurons) is finite because the clustering and pruning algorithms do not let m_1 become infinity. The number of neurons m_1 is changed by the clustering and pruning algorithms, and m_1 only changes the dimension of B_{k-1}^T and $\theta(k-1)$; thus the stability result is preserved.

7 Proposed Algorithm

The proposed algorithm is finally as follows:

1. Select the following parameters: for the clustering algorithm as $0 < r < 1 \in \Re$ and for the pruning algorithm as $d_u \in N$ $(L = L + \Delta L, \Delta L = 5d_u)$. If r is bigger, more neurons could be generated. If d_u is smaller, more neurons could be pruned. If there are many neurons that are generated and pruned, then it could cause like a chattering in the modeling. Consequently, only the required neurons should be generated and pruned in the algorithm.
2. For the first data $k = 1$ (where k is the iterations number) and $m_1 = 1$ (where m_1 is the hidden neurons number), the initial parameters of the least square algorithm are $P_1 \in \Re^{3(m_2+m_1)\times 3(m_2+m_1)}$ with diagonal elements, $v_{11}(1) = y(1)$, $c_{11}(1) = x(1)$, and $w_{11}(1) = \text{rand} \in (5, 15)$ (v_{11} is the initial parameter of the consequent part, and c_{11} and w_{11} are the centers and widths of the membership function of the antecedent part), rand is a random number which lets to find some similar alternative results, $v_{21}(1) = y(1)$, $c_{21}(1) = x_n(1)$, $w_{21}(1) = \text{rand}$, m_2 =size of the input $u(k)$, and the initial parameter of the clustering and pruning algorithm is $d_1(1) = 1$ (where d is the density parameter).
3. For the other data $k \geq 2$, evaluate the network parameters $z_{1j}(k-1)$, $b_1(k-1)$, $z_{2j}(k-1)$, and $b_2(k-1)$ with (8.6), evaluate the output of the network $\widehat{y}(k-1)$ with (8.7), (8.8), and (8.18), evaluate the modeling error $e(k-1)$ with (8.18), update the parameters of the least square algorithm $v_{1j}(k)$, $c_{1j}(k)$, $w_{1j}(k)$, $v_{2j}(k)$, $c_{2j}(k)$, and $w_{2j}(k)$ with (8.30) (where $j = 1 \ldots m_1$ for $\widehat{f_{k-1}}$ and $j = 1 \ldots m_2$ for $\widehat{g}_{k-1}$), and evaluate the parameter of the clustering and pruning algorithms $p(k-1)$ with (8.21).
   The updating of the clustering algorithm is as follows:
4. If $p(k-1) < r$, then a new neuron is generated ($m_1 = m_1 + 1$), where $r \in (0, 1)$ (i.e., the number of neuron is increased by one), assign initial values to the new neuron as $c_{1m_1+1}(k)$, $w_{1m_1+1}(k)$, $v_{1m_1+1}(k)$, and $d_{m_1+1}(k)$ with (8.22), the values are assigned for $P_k \in \Re^{3(m_2+m_1+1)\times 3(m_2+m_1+1)}$ from elements $m_2 + 1$ to $m_2 + m_1 + 1$ with diagonal elements (where P_k, $v_{1j}(k)$, $c_{1j}(k)$, and $w_{1j}(k)$ are the parameters of the least square algorithm, and $d_j(k)$ is the density parameter, $j = 1 \ldots m_1$), and go to 3.
5. If $p(k-1) \geq r$, then a neuron is not generated, and if $z_{1j}(k-1) = p(k-1)$, the winner neuron j^* is obtained, the value of the density $d_{j^*}(k)$ of this hidden neuron is updated with (8.23), the winner neuron is a hidden neuron that increments its importance in the algorithm, and go to 3.
   The updating of the pruning algorithm is as follows:
6. For the case that $k = L$, the pruning algorithm works (the pruning algorithm does not work in each iteration), evaluate the minimum density $d_{\min}(k)$ with (8.24), and L is updated as $L = L + \Delta L$.
7. If $m_1 \geq 2$ and if $d_{\min}(k) \leq d_u$, this hidden neuron is pruned, where $d_u \in N$ is the density umbral, if $d_j(k-1) = d_{\min}(k)$ the looser neuron j_* is obtained, the looser neuron is the least important neuron of the algorithm, assign values

to $c_{1j}(k)$, $w_{1j}(k)$, $v_{1j}(k)$, and $d_j(k)$ with (8.25) and (8.26) to prune the looser neuron j_*, assign values for $P_k \in \Re^{3(m_2+m_1-1)x3(m_2+m_1-1)}$ from elements m_2+1 to m_2+m_1-1 with diagonal elements to prune the looser neuron j_*, (where P_k, $v_{1j}(k)c_{1j}(k)$, and $w_{1j}(k)$ are the parameters of the least square algorithm and $d_j(k)$ is the density parameter $j = 1 \ldots m_1$), update m_1 as $m_1 = m_1 - 1$ (i.e., the number of hidden neurons is decreased by one), and go to 3.

8. If $d_{\min}(k) > d_u$ or $m_1 = 1$, this neuron is not pruned, and go to 3.

8 Simulations

In this section, the suggested online evolving intelligent system is applied for the modeling of nonlinear system with dead-zone input. Note that the structure and parameters updating of the proposed approach work at the same time. The algorithm of this chapter is compared with the Kalman filter algorithm of [25, 27, 35] and with the evolving algorithm of [36] because the above neuro fuzzy systems have a similar structure. In this section the proposed algorithm is called ModifiedEvolving, the Kalman filter is called KalmanFilter, and the evolving algorithm is called Evolving.

The root mean square error (RMSE) is used to obtain the algorithms' performance, and it is given as follows [18, 22]:

$$\text{RMSE} = \left(\frac{1}{N} \sum_{k=1}^{n} \widetilde{x}_i^2(k) \right)^{\frac{1}{2}}, \tag{8.32}$$

where $\widetilde{x}_i(k)$ is the state error of (8.20), and n is the state number.

Example 8.1 The nonlinear system used for the modeling is expressed as follows:

$$\begin{aligned} x_1(k) &= x_1(k-1) + T x_2(k-1), \\ x_2(k) &= x_2(k-1) \\ &+T\left[1.3\left(\frac{1-\exp(-x_1(k-1))}{1+\exp(-x_2(k-1))}\right) - 2\left(\frac{\sin(x_1(k-1))\cos(x_2(k-1))}{x_1^2(k-1)+x_2^2(k-1)+1}\right)\right. \\ &\left.+\left(\frac{\sin(x_1(k-1)x_2(k-1))+\cos^2(x_2(k-1))+1.5}{1.5}\right)u(k-1) + 0.2\,\text{rand}\right], \end{aligned} \tag{8.33}$$

where $v(k-1) = 0.18\sin(1.5\pi(k-1)T) + 0.28\sin(0.5\pi(k-1)T)$, $T = 0.01$ is the system input, the dead-zone $u(k-1)$ is given as (8.2) with $m_r = 0.1$, $m_l = 0.1$, $b_r = 0.1$, $b_l = -0.1$ for the first half of the time, and $m_r = 0.2$, $m_l = 0.2$, $b_r = 0.2$, $b_l = -0.2$ for the second half of the time. $x_1(1) = 0.5$, $x_2(1) = 0$ are the initial conditions. The nonlinear system has the form (8.1). The data for 2000 iterations are used for the modeling. The signal 0.2 rand is a noise signal where rand are random numbers.

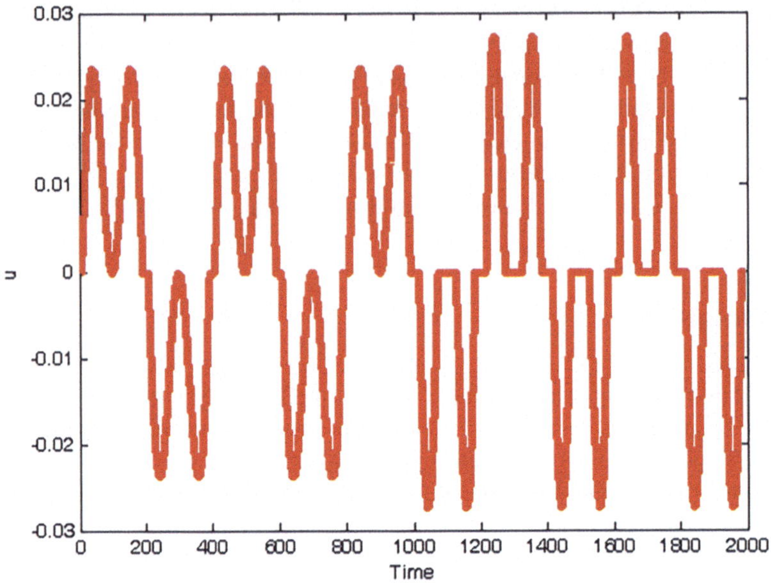

Fig. 8.3 The dead-zone

KalmanFilter is given in [25, 27, 35], with parameters $x(k) = [x_1(k), x_1(k)]^T$, $\widehat{x}(k) \in \Re^2$, $P_{11} = P_{21} = diag(1 \times 10^{-6}) \in \Re^{2(1+1)\times 2(1+1)}$.

Evolving is given in [36] with parameters $x(k) = [x_1(k), x_1(k)]^T, \widehat{x}(k) \in \Re^2$, $S = diag(0.1) \in \Re^{2\times 2}$, $P_{11} = P_{21} = diag(100) \in \Re^{3(1+1)\times 2(1+1)}$, $r = 0.7$, and $d_u = 4$.

ModifiedEvolving is given as (8.4), (8.6) or (8.4), (8.7), (8.8) with parameters $x(k) = [x_1(k), x_2(k)]^T, \widehat{x}(k) \in \Re^2$, $s = 0.1$, $P_1 = diag(100) \in \Re^{3(1+1)\times 2(1+1)}$, $r = 0.7$, and $d_u = 4$.

Figure 8.3 shows the dead-zone. The states' approximation is shown in Figs. 8.4 and 8.5, and state errors are shown in Fig. 8.5. The growth of the hidden neurons is shown in Fig. 8.6. Table 8.1 shows the comparison of the RMSE and neurons for the modeling of three algorithms.

From Fig. 8.3, it is shown that the dead-zone changes in the half of the time. From Figs. 8.4, 8.5, 8.6, and 8.7 and Table 8.1, it can be seen that ModifiedEvolving achieves better accuracy when compared with both the Evolving and KalmanFilter because the first follows better the signals than the others, and also the RMSE and neuron number for the first are smaller than for the others. Consequently, the proposed algorithm is good for the modeling of the first nonlinear system with dead-zone input.

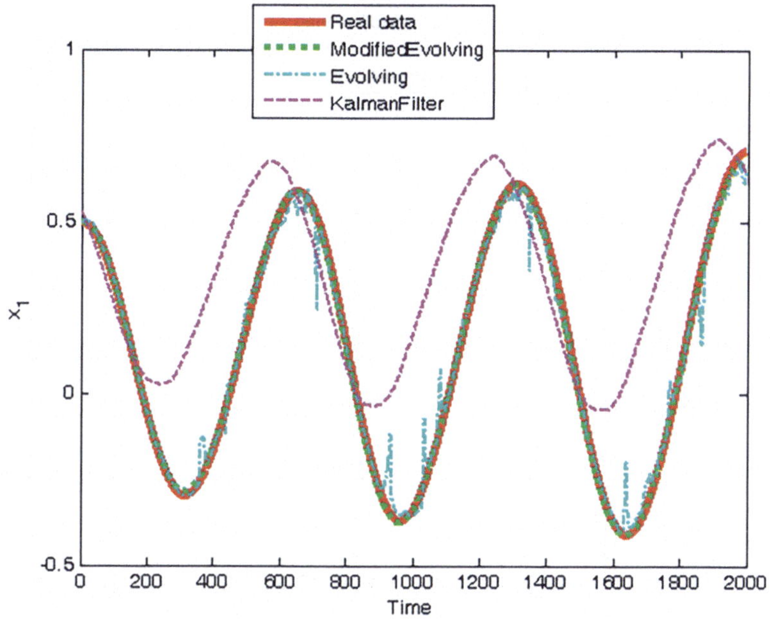

Fig. 8.4 The approximation of the state x_1

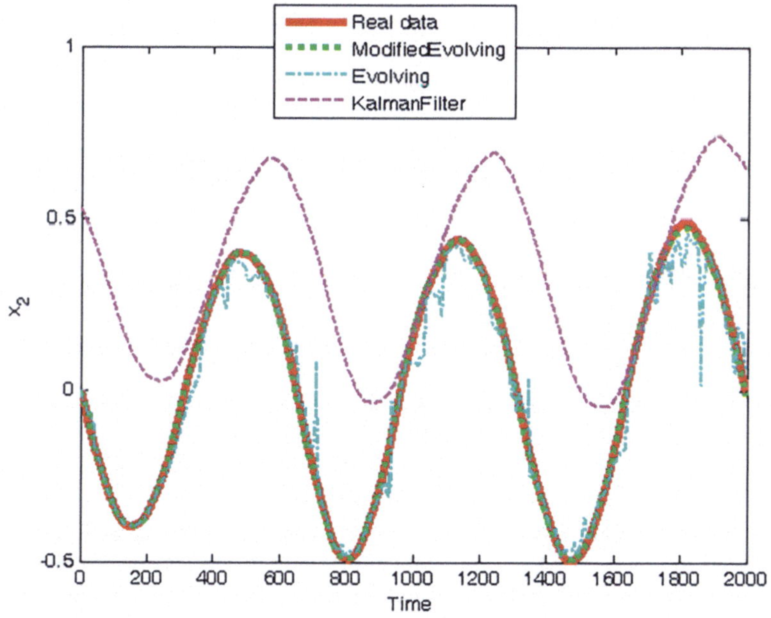

Fig. 8.5 The approximation of the state x_2

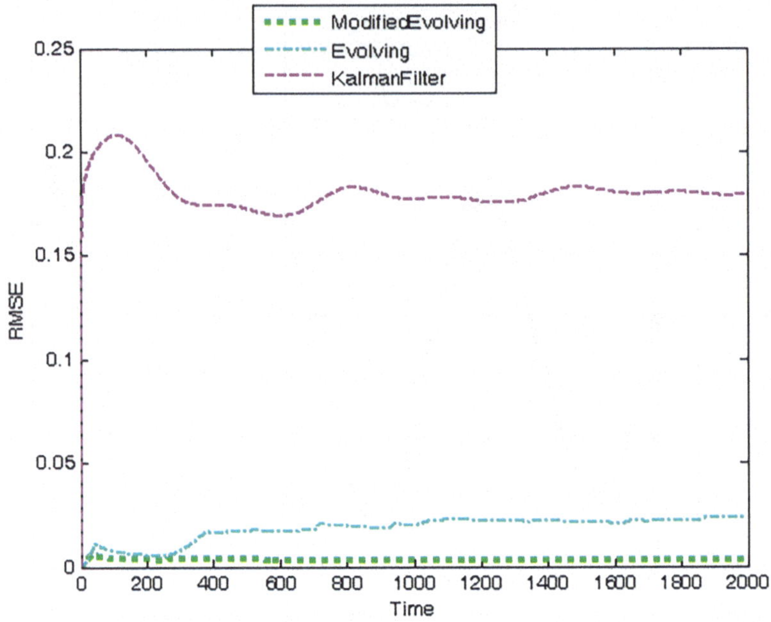

Fig. 8.6 The RMSE

Table 8.1 Results for Example 8.1

Methods	Neurons	RMSE
KalmanFilter	12	0.1798
Evolving	12	0.0240
ModifiedEvolving	7	0.0036

Example 8.2 The nonlinear system used for the modeling is expressed as follows [3]:

$$
\begin{aligned}
&x_1(k) = x_1(k-1) + Tx_2(k-1),\\
&x_2(k) = x_2(k-1)\\
&+T\Bigg[-2.3\left(\tfrac{1-\exp(-x_1(k-1))}{1+\exp(-x_2(k-1))}\right) + 3.7\left(\tfrac{x_2(k-1)\sin(x_1(k-1)x_2(k-1))\cos(x_2(k-1))}{x_1^2(k-1)+x_2^2(k-1)+1}\right)\\
&+1.5x_1(k-1)x_2(k-1) + 0.7x_1(k-1)x_2^3(k-1)\sin(2x_1(k-1))\\
&+0.4x_1^2(k-1)x_2(k-1) + 3.5u(k-1) - 0.5\,\mathrm{rand}\Bigg],
\end{aligned}
\tag{8.34}
$$

where $v(k-1) = 0.18\sin(1.5\pi(k-1)T) + 0.28\sin(0.5\pi(k-1)T)$, $T = 0.01$ is the system input, the dead-zone $u(k-1)$ is given as (8.2) with $m_r = 0.1$, $m_l = 0.1$, $b_r = 0.1$, $b_l = -0.1$ for the first half of the time, and $m_r = 0.05$, $m_l = 0.05$, $b_r = 0.05$, $b_l = -0.05$ for the second half of the time. $x_1(1) = 0.5$, and $x_2(1) = 0$ are the initial conditions. The nonlinear system has the form (8.1). The data for 2000

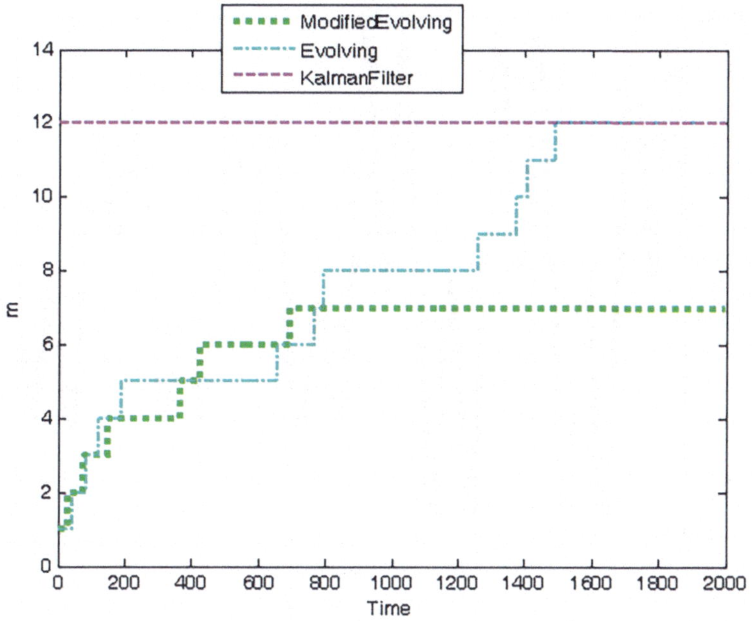

Fig. 8.7 The neurons number

iterations are used for the modeling. The signal -0.5 rand is a noise signal where rand are random numbers.

KalmanFilter is given in [25, 27, 35], with parameters $x(k) = [x_1(k), x_1(k)]^T$, $\widehat{x}(k) \in \Re^2$, $P_{11} = P_{21} = diag(1 \times 10^{-6}) \in \Re^{2(1+1)\times 2(1+1)}$.

Evolving is given in [36] with parameters $x(k) = [x_1(k), x_1(k)]^T, \widehat{x}(k) \in \Re^2$, $S = diag(0.1) \in \Re^{2\times 2}$, $P_{11} = P_{21} = diag(100) \in \Re^{3(1+1)\times 2(1+1)}$, $r = 0.7$, and $d_u = 4$.

ModifiedEvolving is given as (8.4), (8.6) or (8.4), (8.7), and (8.8) with parameters $x(k) = [x_1(k), x_2(k)]^T, \widehat{x}(k) \in \Re^2$, $s = 0.1$, $P_1 = diag(100) \in \Re^{3(1+1)\times 3(1+1)}$, $r = 0.7$, and $d_u = 4$.

Figure 8.8 shows the dead-zone. The states' approximation is shown in Figs. 8.9 and 8.10, and state errors are shown in Fig. 8.11. The growth of the hidden neurons is shown in Fig. 8.12. Table 8.2 shows the comparison of the RMSE and neurons for the modeling of three algorithms.

From Fig. 8.8, it is shown that the dead-zone changes in the half of the time. From Figs. 8.9, 8.10, 8.11, and 8.12 and Table 8.2, it can be seen that ModifiedEvolving achieves better accuracy when compared with both the Evolving and KalmanFilter because the first follows better the signals than the others, and also the RMSE and neuron number for the first are smaller than for the others. Consequently, the proposed algorithm is good for the modeling of the second nonlinear system with dead-zone input.

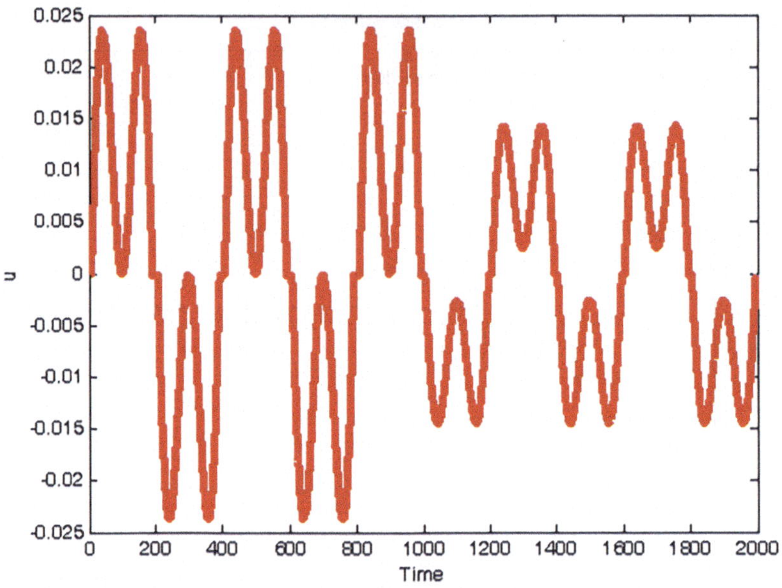

Fig. 8.8 The dead-zone

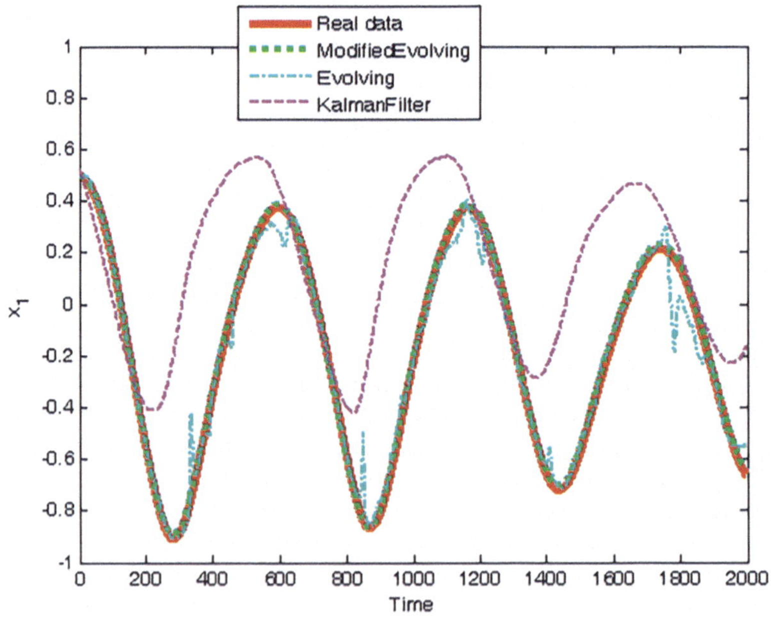

Fig. 8.9 The approximation of the state x_1

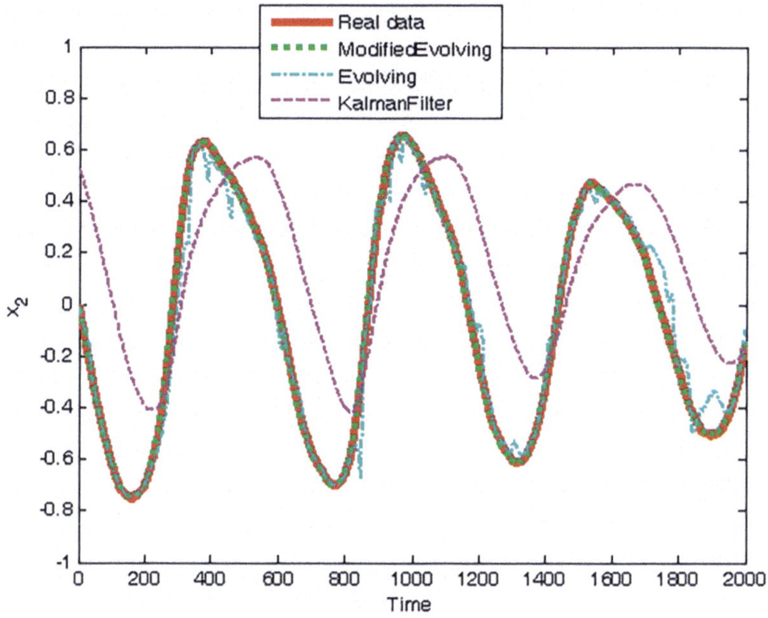

Fig. 8.10 The approximation of the state x_2

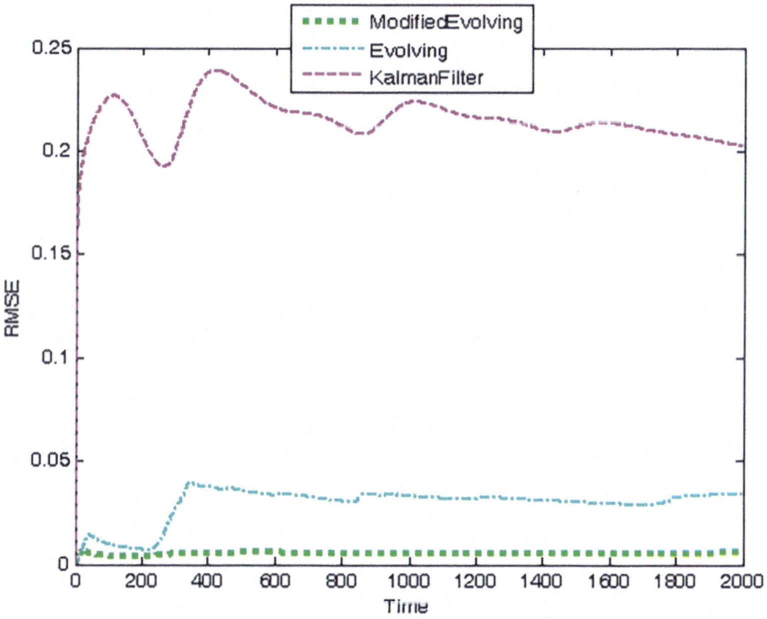

Fig. 8.11 The RMSE

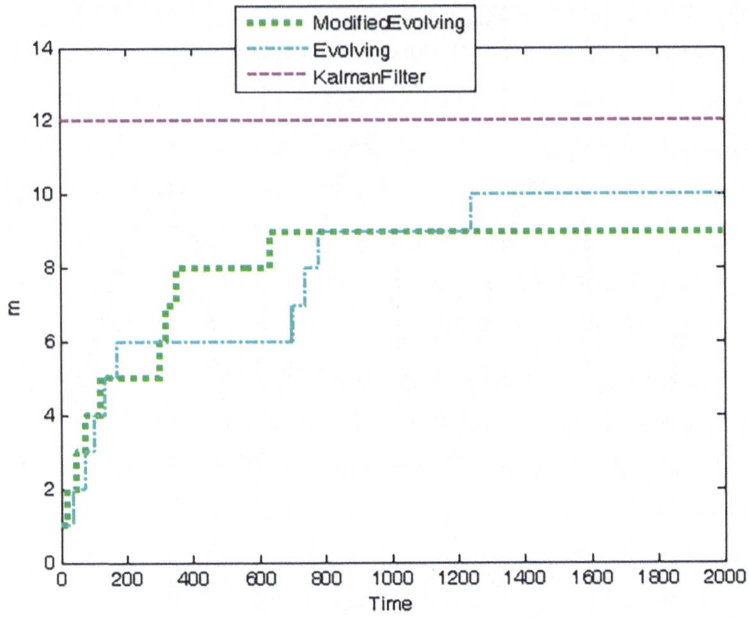

Fig. 8.12 The neurons number

Table 8.2 Results for Example 8.2

Methods	Neurons	RMSE
KalmanFilter	12	0.2030
Evolving	10	0.0337
ModifiedEvolving	9	0.0059

Remark 8.8 The nonlinear systems (8.33) and (8.34) are different because in the first f_{k-1} is bounded and g_{k-1} changes with the time, while in the second f_{k-1} is not bounded and g_{k-1} is constant. In addition, the noise signals are different for both models.

Remark 8.9 The proposed algorithm approximates the behavior of the nonlinear systems (8.33), (8.34) which includes f_{k-1} and g_{k-1}, and the dead-zone $u(k-1)$ is inside of g_{k-1}; therefore, the proposed algorithm approximates the behavior of the nonlinear systems considering the dead-zone behavior.

9 Concluding Remarks

In this chapter, an approach using an evolving intelligent system was presented for the modeling of nonlinear systems with dead-zone input. It is effective, as it does not require retraining of the whole model. It is based on recursive building of the hidden neuron base by unsupervised and supervised learning, the hidden neuron-based model structure learning and parameter estimation. The simulation showed the proposed evolving system achieves better performance when compared with both the Kalman filter and evolving algorithms for the modeling of nonlinear systems with dead-zone input. The results illustrate the viability, efficiency, and the potential of the approach when it used a limited amount of initial information, especially important in autonomous systems and robotics. As a future research, the proposed evolving intelligent system will be applied for the modeling of robotic or mechatronic systems.

References

1. S. Abrir, W.F. Xie, C.Y. Su, Adaptive tracking of nonlinear systems with non-symmetric dead-zone input. Automatica **43**, 522–530 (2007)
2. Y.J. Liu, N. Zhou, Observer-based adaptive fuzzy-neural control for a class of uncertain nonlinear systems with unknown dead-zone input. ISA Trans. **49**, 462–469 (2010)
3. J.H. Perez Cruz, E. Ruiz Velazquez, J.J. Rubio, C.A. de Alva Padilla, Robust adaptive neurocontrol of SISO nonlinear systems preceded by unknow deadzone. Math. Prob. Eng. **2012**, 1–22 (2012)
4. J.H. Perez Cruz, J.J. Rubio, E. Ruiz-Velazquez, G. Solis-Perales, Tracking control based on recurrent neural networks for nonlinear systems with multiple inputs and unknown deadzone. Abstr. Appl. Anal. **2012**, 1–18 (2012)
5. T. Zhang, S.S. Ge, Adaptive neural network tracking control of MIMO nonlinear systems with unknown dead zones and control directions. IEEE Trans. Neural Networks **20**(3), 483–497 (2009)
6. P. Angelov, E. Lughofer, X. Zou, Evolving fuzzy classifiers using different model architectures. Fuzzy Sets Syst. **159**(23), 3160–3182 (2008)
7. P. Angelov, Fuzzily connected multimodel systems evolving autonomously from data streams. IEEE Trans. Syst. Man Cybern. Part B: Cybern. **41**(4), 898–910 (2011)
8. A. Bouchachia, Incremental learning with multi-level adaptation. Neurocomputing **74**(11), 1785–1799 (2011)
9. C.F. Juang, T.C. Chen, W.Y. Cheng, Speedup of implementing fuzzy neural networks with high-dimensional inputs through parallel processing on graphic processing units. IEEE Trans. Fuzzy Syst. **19**(4), 717–728 (2011)
10. D. Leite, R. Ballini, P. Costa, F. Gomide, Evolving fuzzy granular modeling from nonstationary fuzzy data streams. Evol. Syst. **3**(2), 65–79 (2012)
11. A. Lemos, W. Caminhas, F. Gomide, Multivariable Gaussian evolving fuzzy modeling system. IEEE Trans. Fuzzy Syst. **19**(1), 91–104 (2011)
12. E. Lughofer, P. Angelov, Handling drifts and shifts in online data streams with evolving fuzzy systems. Appl. Soft Comput. **11**(2), 2057–2068 (2011)
13. E. Lughofer, Single pass active learning with conflict and ignorance. Evol. Syst. **3**, 251–271 (2012)
14. E. Lughofer, A dynamic split-and-merge approach for evolving cluster models. Evol. Syst. **3**, 135–151 (2012)
15. L. Maciel, A. Lemos, F. Gomide, R. Ballini, Evolving fuzzy systems for pricing fixed income options. Evol. Syst. **3**, 5–18 (2012)
16. H.J. Rong, N. Sundararajan, G.B. Huang, P. Saratchandran, Sequential adaptive fuzzy inference system (SAFIS) for nonlinear system identification and prediction. Fuzzy Sets Syst. **157**(9), 1260–1275 (2006)

17. H.J. Rong, N. Sundararajan, G.B. Huang, G.S. Zhao, Extended sequential adaptive fuzzy inference system for classification problems. Evol. Syst. **2**(2), 71–82 (2011)
18. J.J. Rubio, D.M. Vazquez, J. Pacheco, Backpropagation to train an evolving radial basis function neural network. Evol. Syst. **1**(3), 173–180 (2010)
19. C.K. Ahn, Takaji-Sugeno fuzzy Hopfield neural networks for h∞ nonlinear system identification. Neural Proces. Lett. **34**(1), 59–70 (2011)
20. C.K. Ahn, Exponential Hw stable learning method for Takaji-Sugeno fuzzy delayed networks. Comput. Math. Appl. **63**(5), 887–895 (2011)
21. X. Ren, X. Lv, Identification of extended Hammerstein systems using dynamic self-optimizing neural networks. IEEE Trans. Neural Networks **22**(8), 1169–1179 (2011)
22. J.J. Rubio, P. Angelov, J. Pacheco, An uniformly stable backpropagation algorithm to train a feedforward neural network. IEEE Trans. Neural Networks **22**(3), 356–366 (2011)
23. J.J. Rubio, M. Figueroa, J.H. Perez Cruz, J. Rumbo, Control to stabilize and mitigate disturbances in a rotatory inverted pendulum. Mex. J. Phys. E **58**(2), 107–112 (2012)
24. J.J. Rubio, Modified optimal control with a backpropagation network for robotic arms. IET Control Theory Appl. **6**(14), 2216–2225 (2012)
25. X. Wang, Y. Huang, Convergence study in a extended Kalman filter-based training of recurrent neural networks. IEEE Trans. Neural Networks **22**(4), 588–600 (2011)
26. Q. Yang, S. Jagannathan, Reinforcement learning controller design for affine nonlinear discrete-time systems using online approximators. IEEE Trans. Syst. Man Cybern. Part B: Cybern. **42**(2), 377–390 (2012)
27. W. Yu, J.J. Rubio, Recurrent neural network training with stable bounding ellipsoid algorithm. IEEE Trans. Neural Networks **20**(6), 983–991 (2009)
28. H. Zhang, W. Wu, M. Yao, Boundedness and convergence of bath back-propagation algorithm with penalty for feedforward neural networks. Neurocomputing **15**, 141–146 (2012)
29. J.J. Rubio, J. Serrano, M. Figueroa, C.F. Aguilar-Ibañez, Dynamic model with sensor and actuator for an articulated robotic arm. Neural Comput. Appl. **24**, 573–581 (2014)
30. J.J. Rubio, J. Pacheco, J.H. Perez-Cruz, F. Torres, Mathematical model with sensor and actuator for a transelevator. Neural Comput. Appl. **24**, 277–285 (2014)
31. S. Haykin, *Neural Networks-A Comprehensive Foundation* (Macmillan, New York, 1994)
32. K.S. Narendra, K. Parthasarathy, Identification and control of dynamic systems using neural networks. IEEE Trans. Neural Networks **1**(1), 4–27 (1990)
33. J.S.R. Jang, C.T. Sun, E. Mizutani, *Neuro-Fuzzy and Soft Computing* (Prentice Hall, Hoboken, 1996)
34. L.X. Wang, *A Course in Fuzzy Systems and Control* (Prentice Hall, Englewood Cliffs, 1997)
35. J.J. Rubio, W. Yu, Nonlinear system identification with recurrent neural networks and dead-zone Kalman filter algorithm. Neurocomputing **70**(13), 2460–2466 (2007)
36. J.J. Rubio, Stability analysis for an online evolving neuro-fuzzy recurrent network in *Evolving Intelligent Systems: Methodology and Applications*, ed. by P. Angelov, D. Filev, N. Kasabov (John Wiley & Sons, New York, 2010), pp. 173–200
37. E.B. Kosmatopoulos, M.M. Polycarpou, M.A. Christodoulou, P.A. Ioannou, High-order neural network structures for identification of dynamic systems. IEEE Trans. Neural Networks **6**(2), 422–431 (1995)
38. J.J. Rubio, W. Yu, Stability analysis of nonlinear systems identification via delayed neural networks. IEEE Trans. Circuits Syst. Part II **54**(2), 161–165 (2007)
39. J.J. Rubio, J.H. Perez Cruz, Evolving intelligent system for the modelling of nonlinear systems with dead-zone input. Appl. Soft Comput. **14**(Part B), 289–304 (2014)

Chapter 9
Evolving Intelligent Algorithms for the Modeling of Brain and Eye Signals

1 Introduction

In recent years, there are two important topics that are related with the modeling; they are the evolving intelligent systems and stable intelligent systems.

The evolving intelligent systems are characterized by abilities to adjust their structure and parameters to the varying characteristics of the environment (with the term of environment embracing processes/phenomena with which the system has to interact and or deal with the users using the system) [1–3]. Some important results are given by Garcia-Cuesta and Iglesias [4], Juang et al. [5], Leite et al. [6], Lemos et al. [7, 8], Lughofer [9], Lughofer and Angelov [10], Lughofer and Bouchot [11], Lughofer [12, 13], Maciel et al. [14], Ordoñez et al. [15], and Rong et al. [16–18]. The problem of the classification of streaming data from a dimensionality reduction perspective is addressed by Garcia-Cuesta and Iglesias [4]. The implementation of a zero-order Takagi-Sugeno-Kang (TSK)-type fuzzy neural network (FNN) is proposed by Juang et al. [5]. An evolving fuzzy granular framework to learn from and model time varying fuzzy input-output data streams is introduced by Leite et al. [6]. A class of evolving fuzzy rule-based system as an approach for multivariable Gaussian adaptive fuzzy modeling is considered by Lemos et al. [7]. A new approach for evolving fuzzy modeling using tree structures is proposed by Lemos et al. [8]. A new algorithm for incremental learning of a specific form of Takagi-Sugeno fuzzy systems is introduced by Lughofer [9]. New approaches to handling drift and shift in online data streams with the help of evolving fuzzy systems (EFSs) are presented by Lughofer and Angelov [10]. In [11], the authors examine approaches for reducing the complexity of EFSs by eliminating local redundancies during training. A new methodology for conducting active learning in a single-pass online learning context is introduced by Lughofer [12]. New dynamic split-and-merge operations for evolving cluster models, which are learned incrementally and

J. de J. Rubio, *Stability Analysis of Neural Networks and Evolving Intelligent Systems*, https://doi.org/10.1007/978-3-031-87282-2_9

expanded on the fly from data streams, are considered by Lughofer [13]. In [14], the authors address option pricing using an evolving fuzzy system model and Brazilian interest rate options data. The use of evolving classifiers for activity recognition from sensor readings in ambient assisted living environments is described by Ordoñez et al. [15]. In [16], a Sequential Adaptive Fuzzy Inference System called SAFIS is developed based on the functional equivalence between a radial basis function network and a fuzzy inference system (FIS). The performance evaluation of the recently developed Sequential Adaptive Fuzzy Inference System (SAFIS) algorithm for classification problems is presented by Rong et al. [17]. In [18], two adaptive fuzzy control schemes including indirect and direct frameworks are developed for suppressing the wing-rock motion. The above systems are evolving and soft; however, they are not guaranteed to be stable.

The stable intelligent systems are characterized to be systems where some kind of stability is guaranteed, i.e., for bounded inputs in the algorithms, there are also bounded outputs and bounded parameters. Some important results are given by Ahn [19], Ahn and Lim [20], Ren and Lv [21], Rubio et al. [22, 23], Wang and Huang [24], Yu and Rubio [25], and Zhang et al. [26]. In [19], an error passivation approach is used to derive a new passive and exponential filter for switched Hopfield neural networks with time delay and noise disturbance. The model predictive stabilization problem for Takagi-Sugeno (T-S) fuzzy multilayer neural networks with general terminal weighting matrix is investigated by Ahn and Lim [20]. Two stable neural networks are introduced by Ren and Lv [21] and Rubio et al. [22]. The aforementioned studies are stable and soft; nevertheless, they are not evolving.

There is research where evolving and stable characteristics are possible and also combined whenever assuring some sort of convergence to optimality given by Rubio et al. [23], Lughofer [27], and Rubio [28]. The systems are novel because they merge the main characteristics of the above techniques into one algorithm which has the main characteristics to be evolving, soft, and stable. See Fig. 9.1.

This chapter presents the comparison of three intelligent algorithms for the modeling of brain and eye signals. The signals could be applied for the patients who cannot move their bodies; therefore, they could use their brains or their eyes to say what they want or need. The algorithms are the SAFIS algorithm [16] which is an evolving intelligent system, SBP [22], which is a stable intelligent system, and SOFMLS [28], which is an evolving and stable intelligent system.

The chapter is organized as follows. In Sect. 2, the SAFIS, SBP, and SOFMLS algorithms are detailed. In Sect. 3, the encephalography (EEG) and electrooculogram (EOG) signals are described. In Sect. 4, the comparison of three algorithms for the modeling of brain and eye signals is presented. Section 5 presents conclusions and suggests future research directions.

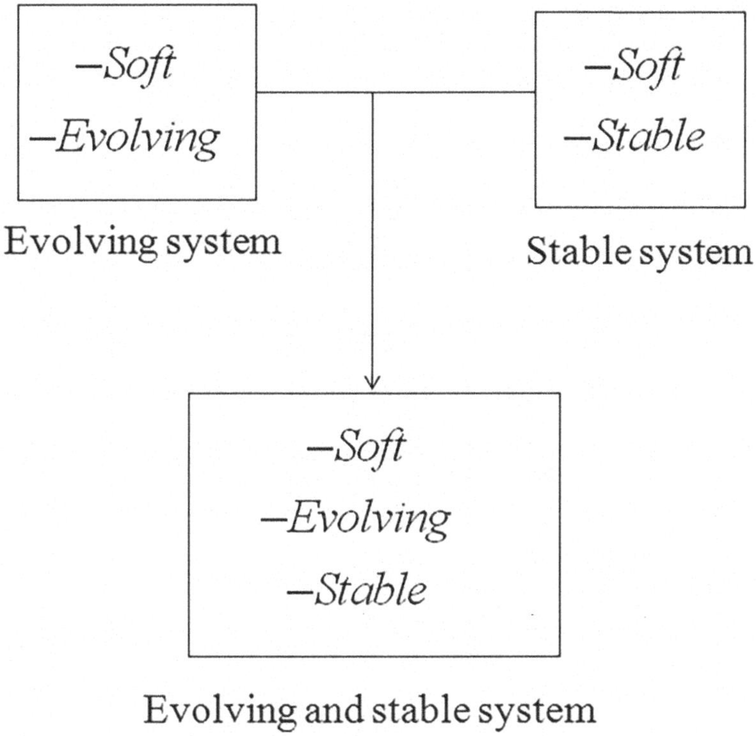

Fig. 9.1 Evolving and stable intelligent systems

2 Preliminaries

In this section the three algorithms of this chapter are described.

2.1 SAFIS Algorithm

A sequential adaptive fuzzy inference system (SAFIS) is developed based on the functional equivalence between a radial basis function network and a fuzzy inference system (FIS). In SAFIS, the concept of "Influence" of a fuzzy rule is introduced, and using this the fuzzy rules are added or removed based on the input data received so far. If the input data do not warrant adding of fuzzy rules, then only the parameters of the "closest" (in a Euclidean sense) rule are updated using an extended Kalman filter (EKF) scheme.

The SAFIS algorithm is summarized as below [16]:

Given the growing and pruning thresholds e_g, e_p for each observation (x_k, y_k), where $x_k \in \Re^{N_x}$, $y_k \in \Re^{N_y}$ and $k = 1, 2, \ldots,$ do:

(1) Compute the overall system output:

$$\widehat{y}_k = \frac{\sum_{n=1}^{N_h} a_n R_n(x_k)}{\sum_{n=1}^{N_h} R_n(x_k)}, \tag{9.1}$$

where

$$R_n(x_k) = \exp\left(-\frac{1}{\sigma_n^2}\|x_k - \mu_n\|^2\right),$$

where N_h is the number of fuzzy rules.

(2) Calculate the parameters required in the growth criterion:

$$\varepsilon_k = \max\left\{\varepsilon_{\max}\gamma^n, \varepsilon_{\min}\right\}, 0 < \gamma < 1, \tag{9.2}$$

$$e_k = y_k - \widehat{y}_k. \tag{9.3}$$

(3) Apply the criterion for adding rules:

If

$$\|x_k - \mu_{rn}\| > \varepsilon_k, \tag{9.4}$$

and

$$E_{\inf}(N_h + 1) = |e_k| \frac{(1.8n\|x_k - \mu_{rn}\|)^{N_x}}{\sum_{n=1}^{N_h+1}(1.8\sigma_n)^{Nx}} > e_g, \tag{9.5}$$

and allocate a new rule with

$$\begin{aligned} a_{Nn+1} &= e_k, \\ \mu_{Nh+1} &= x_k, \\ \sigma_{Nh+1} &= n\|x_k - \mu_{rn}\|. \end{aligned} \tag{9.6}$$

Else, adjust the system parameters $a_{rn}, \mu_{rn}, \sigma_{rn}$ for the nearest rule only by using the EKF method:

$$\begin{aligned} K_k &= P_{k-1} B_k \left[R_k + B_k^T P_{k-1} B_k\right]^{-1}, \\ \theta_k &= \theta_{k-1} + K_k e_k, \\ P_k &= \left[I - K_k B_k^T\right] P_{k-1} + qI, \end{aligned} \tag{9.7}$$

where $\theta_k = [\theta_1 \cdots \theta_{rn} \cdots \theta_{Nh}]^T = [a_1, \mu_1, \sigma_1, \ldots, a_{rn}, \mu_{rn}, \sigma_{rn}, \ldots, a_{Nh}, \mu_{Nh}, \sigma_{Nh}]$.
Check the criterion for pruning the rule:
If

$$E_{\inf}(rn) = |a_{rn}| \frac{(1.8\sigma_{rn})^{N_x}}{\sum\limits_{n=1}^{N_h+1} (1.8\sigma_n)^{Nx}}, \tag{9.8}$$

remove the rnth rule, and reduce the dimensionality of EKF.
EndIf. EndIf.

Remark 9.1 The significance of a neuron proposed in GAP–RBF is defined based on the average contribution of an individual neuron to the output of the RBF network. Under this definition, one may need to estimate the input distribution range $S(X) = \frac{|a_{rn}|}{\sum\limits_{n=1}^{N_h+1} (1.8\sigma_n)^{Nx}}$. However, the influence of a rule introduced in this chapter is different from the significance of a neuron proposed in GAP–RBF. In fact, the influence of a neuron is defined as the relevant significance of the neuron compared to summation of significance of all the existing RBF neurons. As seen from Eq. (9.8), with the introduction of influence one need not estimate the input distribution range and the implementation has been simplified.

Remark 9.2 In parameter modification, SAFIS utilizes a winner rule strategy similar to the work done by Huang et al. The key idea of the winner rule strategy is that only the parameters related to the selected winner rule are updated by the EKF algorithm in every step. The "winner rule" is defined as the rule that is the closest (in the Euclidean distance sense) to the current input data as in. As a result, in SAFIS, a fast computation is achieved.

Remark 9.3 In SAFIS, some parameters need to be decided in advance according to the problems considered. They include the distance thresholds ($\varepsilon_{\max}, \varepsilon_{\min}, \gamma$), the overlap factor ($n$) for determining the width of the newly added rule, the growing threshold (e_g) for a new rule, and the pruning threshold (e_p) for removing an insignificant rule. A general selection procedure for the predefined parameters is given as follows: max is set to around the upper bound of input variables, $\varepsilon_{\min}$ is set to around 10% of $\varepsilon_{\max}$, and γ is set to around 0.99. e_p is set to around 10%

of e_g. The overlap factor (n) is utilized to initialize the width of the newly added rule and chosen according to different problems. It is suggested to be chosen in the range [1.0, 2.0]. The growing threshold e_g is chosen according to the system performance. The smaller the e_g, the better the system performance, but the resulting system structure is more complex.

2.2 SBP Algorithm

The stable backpropagation (SBP) algorithm is developed with a new time varying rate to guarantee its uniformly stability for online identification, and its identification error converges to a small zone bounded by the uncertainty. The weights' error is bounded by the initial weights' error, i.e., overfitting is eliminated in the mentioned algorithm [29].

The SBP algorithm is as follows [22]:

(1) Obtain the output of the nonlinear system $y(k)$ with Eq. (9.9). Note that the nonlinear system may have the structure represented by Eq. (9.9); the parameter N is selected according to this nonlinear system.

$$y(k) = f\left[X_k\right], \tag{9.9}$$

where $X_k = [x_1(k)\ldots, x_i(k), \ldots, x_N(k)]^T = [y(k-1), \ldots, y(k-n), u\,(k-1)\,, \ldots, u\,(k-m)]^T \in \Re^{N\times 1}$ $(N = n+m)$ is the input vector, $u(k-1) \in \Re$ is the input of the plant, $y(k) \in \Re$ is the output of the plant, and f is an unknown nonlinear function, $f \in C^\infty$.

(2) Select the following parameters: V_1 and W_1 as random numbers between 0 and 1, M as an integer number, and α_0 as a positive value smaller or equal to 1; obtain the output of the NN $\widehat{y}(1)$ with Eq. (9.10).

$$\begin{aligned}
\widehat{y}(k) &= V_k\Phi_k = \sum_{j=1}^{M} V_{jk}\phi_{jk}, \\
\Phi_k &= \left[\phi_{1k}, \ldots, \phi_{jk}, \ldots, \phi_{Mk}\right]^T, \\
\phi_{jk} &= \tanh(\sum_{i=1}^{N} W_{ijk}x_i(k)).
\end{aligned} \tag{9.10}$$

(3) For each iteration k, obtain the output of the NN $\widehat{y}(k)$ with Eq. (9.10), also obtain the identification error $e(k)$ with Eq. (9.11), and update the parameters V_{jk+1} and W_{ijk+1} with Eq. (9.12).

$$e(k) = \widehat{y}(k) - y(k), \tag{9.11}$$

$$\begin{aligned} V_{jk+1} &= V_{jk} - \alpha_k \phi_{jk} e(k), \\ W_{ijk+1} &= W_{ijk} - \alpha_k \sigma_{ijk} e(k), \end{aligned} \tag{9.12}$$

where the new time varying rate α_k is

$$\alpha_k = \frac{\alpha_0}{2\left(\frac{1}{2} + \sum_{j=1}^{M} \phi_{jk}^2 + \sum_{j=1}^{M}\sum_{i=1}^{N} \sigma_{ijk}^2\right)},$$

where $i = 1, \ldots, N$, $j = 1, \ldots, M$, $\sigma_{ijk} = V_{jk}\text{sech}^2(\sum_{i=1}^{N} W_{ijk} x_i(k)) x_i(k) \in \Re$.

Remark 9.4 There are two conditions for applying this algorithm for nonlinear systems: The first one is that the nonlinear system may have the form described by (9.9), and the second one is that the uncertainty $\mu(k)$ may be bounded.

Remark 9.5 The value of the parameter used for the stability of the algorithm $\overline{\mu}$ is unimportant, because this parameter is not used in the algorithm. The bound of $\mu(k)$ is needed to guarantee the stability of the algorithm, but it is not used in the BP algorithm (9.10), (9.11), (9.12).

Remark 9.6 The proposed NN has one hidden layer. It was reported in the literature that a feedforward neural network with one hidden layer is enough to approximate any nonlinear system.

Remark 9.7 Note that the behavior of the algorithm could be improved by changing the values of M or α_0.

2.3 SOFMLS Algorithm

An online self-organizing fuzzy modified least-square (SOFMLS) network has the ability to reorganize the model and adapt itself to a changing environment where both the structure and learning parameters are performed simultaneously. The stability of the mentioned algorithm is guaranteed, and the bound for the average identification error is found.

The SOFMLS algorithm is as follows [28]:

(1) Select the following parameters: The parameter of the modified least square algorithm is $R_2 > 0 \in \Re$, the parameter of the clustering algorithm is $0 < r < 1 \in \Re$, and the parameter of the pruning algorithm is $d_u \in N$ ($L = L + \Delta L$, $\Delta L = 5d_u$).
(2) For the first data $k = 1$ (where k is the number of iterations), $M = 1$ (where M is the number of rules or clusters), the initial parameters of the modified least

square algorithm are $P_1 = cI \in \Re^{3M\times 3M}$ (where $0 < c \in \Re$), $v_1(1) = y(1)$, $c_1(1) = \frac{\sum_{i=1}^{N} x_i(1)}{N}$, and $w_1(1) = \text{rand} \in (0, 1)$ (v_1 is the initial parameter of the consequent part, c_1 and w_1 are the centers and widths of the membership function of the antecedent part), and the initial parameter of the clustering and pruning algorithms is $d_1(1) = 1$ (where d is the density parameter).

(3) For the other data where $k \geq 2$, evaluate the fuzzy network parameters $z_j(k-1)$ and $b(k-1)$ with (9.13), evaluate the output of the fuzzy network $\widehat{y}(k-1)$ with (9.13), (9.14), and (9.15), also evaluate the identification error $e(k-1)$ with (9.16), update the parameters of the modified least square algorithm $v_j(k)$, $c_j(k)$, and $w_j(k)$ with (9.17), and evaluate the parameter of the clustering and pruning algorithm $p(k-1)$ with (9.18).

$$\begin{aligned} b(k-1) &= \sum_{j=1}^{M} z_j(k-1), \\ z_j(k-1) &= \exp\left[-\gamma_j^2(k-1)\right], \\ \gamma_j(k-1) &= \frac{\sum_{i=1}^{N} w_j(k-1)\left(x_i(k-1)-c_j(k-1)\right)}{N}, \end{aligned} \tag{9.13}$$

$$\phi_j(k-1) = z_j(k-1)/b(k-1), \tag{9.14}$$

$$\widehat{y}(k-1) = \sum_{j=1}^{M} v_j(k-1)\phi_j(k-1) = V^T(k-1)\Phi(k-1), \tag{9.15}$$

$$e(k-1) = \widehat{y}(k-1) - y(k-1), \tag{9.16}$$

$$\begin{aligned} \theta(k) &= \theta(k-1) - \tfrac{1}{Q_{k-1}} P_k B_{k-1} e(k-1), \\ P_k &= P_{k-1} - \tfrac{1}{R_{k-1}} P_{k-1} B_{k-1} B_{k-1}^T P_{k-1}, \end{aligned} \tag{9.17}$$

$$p(k-1) = \max_{1\leq j\leq M} z_j(k-1). \tag{9.18}$$

The updating of the clustering algorithm is as follows:

(4) If $p(k-1) \geq r$, then a rule is not generated, the winner rule j^* is presented when $z_j(k-1) = p(k-1)$, and the value of the density $d_{j^*}(k)$ of this rule is updated with (9.19). The winner rule is a rule that increments its importance in the algorithm. Go to 3.

$$d_{j^*}(k) = d_{j^*}(k) + 1. \tag{9.19}$$

(5) If $p(k-1) < r$, then a new rule is generated ($M = M + 1$), where $r \in (0, 1)$ (e.g., the number of rules is increased by 1), the initial values of $c_{M+1}(k)$, $w_{M+1}(k)$, $v_{M+1}(k)$, and $d_{M+1}(k)$ are assigned to the new rule with (9.20), and the missing parameters are added to have $P_k \in \Re^{3(M+1)\times 3(M+1)}$ with diagonal elements (where P_k, $v_j(k)$, $c_j(k)$, and $w_j(k)$ are the parameters of the modified least square algorithm, and $d_j(k)$ is the parameter of the density, $j = 1 \ldots M$). Go to 3.

$$c_{M+1}(k) = \frac{\sum_{i=1}^{N} x_i(k)}{N}, \quad w_{M+1}(k) = \frac{\sum_{i=1}^{N} [x_i(k) - c_{j*}(k)]}{N}, \\ v_{M+1}(k) = y(k), \quad d_{M+1}(k) = 1. \tag{9.20}$$

The updating of the pruning algorithm is as follows:

(6) For the case where $k = L$, the pruning algorithm works (the pruning algorithm is not active at each iteration) and evaluates the minimum density $d_{\min}(k)$ with (9.21), and L is updated as $L = L + \Delta L$.

$$d_{\min}(k) = \min_{1 \le j \le M} d_j(k). \tag{9.21}$$

(7) If $M \ge 2$ and $d_{\min}(k) \le d_u$, then this rule is pruned, where $d_u \in N$ is the density threshold, and the looser rule j_* is presented when $d_j(k) = d_{\min}(k)$. The looser rule is the least important rule of the algorithm, the values of $c_j(k)$, $w_j(k)$, $v_j(k)$, and $d_j(k)$ are assigned with (9.22) and (9.23) to prune the looser rule j_*, and in the same way, the values of P_k are assigned to prune the looser rule j_* (where P_k, $v_j(k)$, $c_j(k)$, and $w_j(k)$ are the parameters of the modified least square algorithm and $d_j(k)$ is the density parameter, $j = 1 \ldots M$), and M is updated as $M = M - 1$ (e.g., the number of rules is decreased by 1). Go to 3.

$$c_{j-1}(k) = c_j(k), \quad w_{j-1}(k) = w_j(k), \\ v_{j-1}(k) = v_j(k), \quad d_{j-1}(k) = d_j(k), \tag{9.22}$$

$$c_M(k) = 0, \quad w_M(k) = 0, \quad v_M(k) = 0, \quad d_M(k) = 0. \tag{9.23}$$

(8) If $d_{\min}(k) > d_u$ or $M = 1$, then this rule is not pruned. Go to 3.

Remark 9.8 The networks of many earlier studies use membership functions, as shown in this study, and they also use the function $\gamma_j(k-1)$. First, in the antecedent part of the networks of the aforementioned references, $2N$ parameters are used for each rule of the multidimensional membership functions, while in the antecedent part of the network used in this study, two parameters are used for each rule of the unidimensional membership functions (9.13). Second, the networks of the aforementioned references use $1/\sigma_{ij}(k-1)$, which can cause singularity in online learning, while the network used in this study uses $w_j(k-1) = 1/\sigma_j(k-1)$ to avoid singularity. Some authors use the sum inference, product inference, or norm inference; however, in this study, the mean inference $\gamma_j(k-1)$ (9.13) is used.

Remark 9.9 The idea to take the maximum of $z_j(k-1)$ as in (9.18) to obtain the winner rule is taken from the competitive learning of the adaptive resonance theory (ART) recurrent neural network (in the case of the ART network, the winner rule is the winner neuron).

Remark 9.10 In an earlier research, the second derivative of an objective function is used to find the unimportant rule. In this study, the density parameter is used to find the unimportant rule. In another study, the density as the population is considered, the population of each cluster is monitored, and if it amounts to less than 1% of the total data samples, the cluster is ignored at this iteration. The rule is ignored as $v_{d\,\min}(k) = 0$, and subsequently, this weight is ignored in the term $\widehat{y}(k-1)$ of (9.15). The cluster is ignored in the algorithm at this iteration, but the rule is not pruned; thus, the network cannot decrease. In other earlier work, two threshold parameters are considered: one for adding rules and the other for removing rules; however, they did not use the density parameter.

Remark 9.11 The parameter M (number of rules) is finite, because the algorithm adds the necessary rules and prunes the unnecessary rules to adapt itself to the changing environment. The number of rules M is changed by the clustering and pruning algorithms, and M changes only the dimension of B_{k-1}^T and $\theta(k-1)$; thus, the stability result is preserved.

Remark 9.12 The value of the parameter used for the stability of the algorithm $\overline{\mu}$ is unimportant, because this parameter is not used in the algorithm. The bound of $\mu(k-1)$ is needed to guarantee the stability in the algorithm.

Remark 9.13 The parameters L and ΔL are needed in (9.21), because the pruning algorithm is not active at each iteration. The initial value of L is ΔL, and the pruning algorithm works at the first time when $k = L$, and consequently, L is increased by ΔL. The pruning algorithm works for each ΔL iteration. The parameter ΔL was determined empirically as $5d_u$; thus, the pruning algorithm has only d_u as the designing parameter. Note that the behavior of the algorithm could be improved by changing the values of c, R_2, r, or d_u.

3 The Brain and Eye Signals

This section describes the characteristics of the brain and eye signals.

3.1 The EEG Signals

The difference of the potential in one membrane is obtained by the exchange between the ions (Na+, Cl-, K+) being in the same. The neurons have a potential difference between the inside and outside which is called rest potential, and this potential represents constant changes because of the impulses given by the neighbor neurons [30, 31]. This potential difference can be measured in the brain cortex using electrodes that convert the ion flow into electric flow. The characteristic of the encephalography signal (EEG) is of 5–300 μV in amplitude and of 0–150 Hz in frequency [32, 33].

The EEG signals are waves similar to periodic, but the waves can change from one time to other, and they have some characteristics which allow the modeling [34, 35], as are the amplitude, the frequency, the morphology, the band, the rhythm, and the duration [30, 33].

The following paragraphs show the characteristics that are considered for an adult in vigilance [30, 33].

Alpha signal. It is the normal rhythm of the bottom and is the most stable and typical in the human. It is found in the frequencies of 8–12 Hz$\pm$ 1 Hz. The amplitude is between 20 and 60 μV. It can be seen generally in posterior regions with more amplitude in the occipital lobes. See Fig. 9.2. It is more evident when the patient is awake with closed eyes and in physical and mental rest, and it is stopped when the eyes are opened or with the mental activity [30, 36].

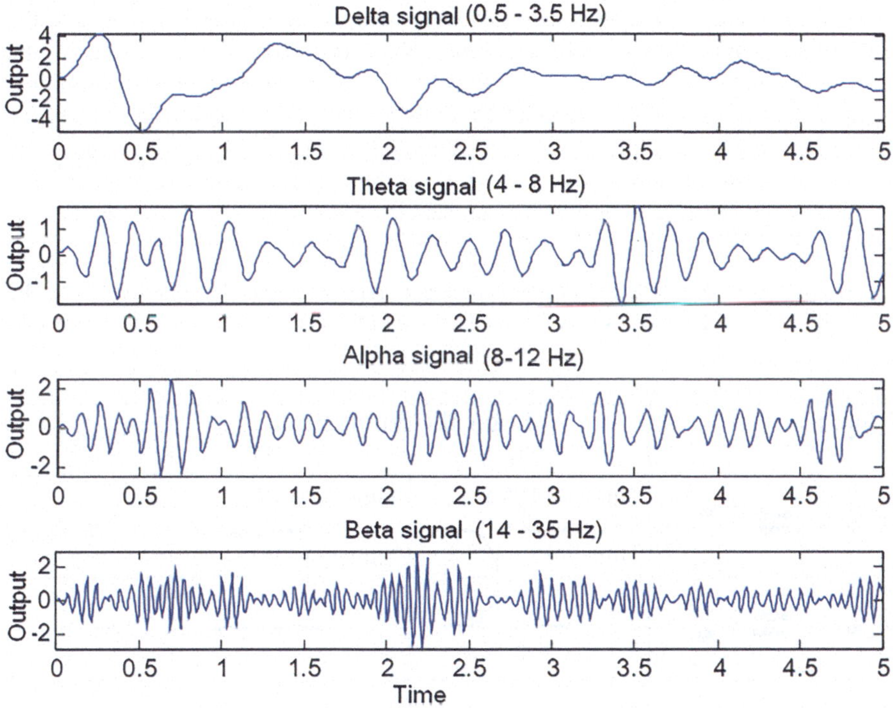

Fig. 9.2 EEG signals

Beta signal. It is found in the frequencies >13 Hz, in general between 14 and 35 Hz. The amplitude is usually low from 5 to 10 μV and is symmetric [30, 36]. See Fig. 9.2.

Theta signal. It has a frequency of 4–8 Hz, is of half of low voltage, and is found in the temporal regions [30, 36]. See Fig. 9.2.

Delta signal. It is found in the second and the third stages of the dream. It has a frequency of 0.5–3.5 Hz, and the amplitude is generally higher than 75 μV [30, 36]. See Fig. 9.2.

3.2 *The EOG Signals*

The EOG signals are the signals obtained as a result of the eye movements of a patient, and these EOG signals are detected using three electrodes: one electrode on the temple, one above and other underneath of the eye. Usually, the detected signals are by direct current (DC) coupling to specify the direction of the gaze. In the experiments of this chapter, three electrodes are placed on the dominant side of the patient eye according to the optimum positions suggested by Hori et al. [37], Rubio et al. [38], and Yamagishi et al. [39].

Figure 9.3 shows the relationship between real eye movements (input) and the EOG signals (output) of the system. Denote the upper and lower thresholds of the vertical channel Ch.V as V1 and V2, respectively, and denote the upper and lower thresholds of the horizontal channel Ch.H as H1 and H2, respectively. When the EOG potential exceeds one of these thresholds, the output assumes ON, and when the EOG potential does not exceed one of these thresholds, the output assumes OFF. The process of transforming the EOG signals from the intention of the patient is as follows [38, 39]:

1. Output Up is when it is obtained an Up behavior: First, Threshold V1 of the vertical channel becomes ON, while Threshold V2 is OFF, and second, Threshold

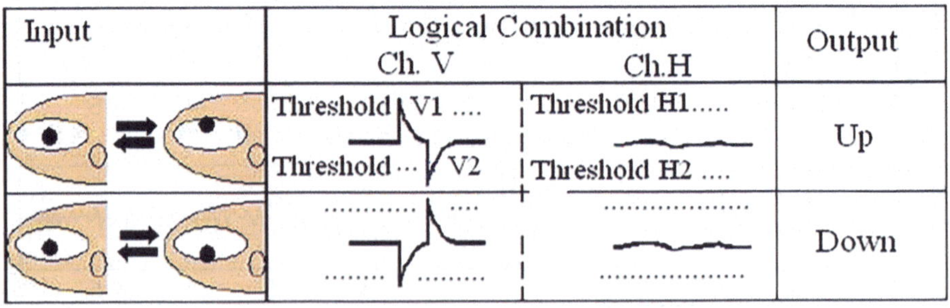

Fig. 9.3 EOG signals

V2 of the vertical channel becomes ON, while Threshold V1 becomes OFF. H1 and H2 of the horizontal channel remain OFF all the time.

2. Output Down is when it is obtained a Down behavior: First, Threshold V2 of the vertical channel becomes ON, while Threshold V1 is OFF, and second, Threshold V1 of the vertical channel becomes ON, while Threshold V2 becomes OFF. H1 and H2 of the horizontal channel remain OFF all the time.

4 Simulations

In this section, the three above detailed algorithms are applied for the modeling of brain and eye signals. The aforementioned signals could be applied for patient who cannot move their bodies; consequently, they could use their brains or their eyes to say what they want or need. The SAFIS of [16], SOFMLS of [28], and SBP of [22] are compared for the modeling of brain signals in Example 1 and for the modeling of eye signals in Example 2. The root mean square error (RMSE) of [22, 28, 33, 40] is used for the comparison results:

$$RMSE = \left(\frac{1}{N} \sum_{k=1}^{N} e^2(k) \right)^{\frac{1}{2}}, \tag{9.24}$$

where $e(k)$ is the learning error of (9.3), (9.16), and (9.11).

4.1 Example 1

Consider real data of brain signals [33] where 5528 pairs $(u(k), y(k))$ of 5.528 s are used for the learning and 1844 pairs $(u(k), y(k))$ for 1.844 s are used for the testing. The alpha signal is obtained in this study because it has more probabilities to be found. The acquisition system is applied with a 28-year-old healthy man when his eyes are closed. The inputs of all the intelligent systems are $y(k)$, $y(k+1)$, $y(k+2)$, $y(k+3)$, and the output of the intelligent systems is $y(k+4)$.

Considering Remark 9.3, the parameters for the SAFIS algorithm [16] are $N_x = 4$, $\gamma = 0.997$, $\varepsilon_{\max} = 2$, $n = 2$, $\varepsilon_{\min} = 0.2$, $e_g = 0.03$, $e_p = 0.003$. Considering Remark 9.7, the parameters of the SBP algorithm [22] are $N = 4$, $M = 4$, $\alpha_0 = 0.25$. Considering Remark 9.13, the parameters of the SOFMLS algorithm [28] are $N = 4$, $P_1 = cI \in \Re^{3x3}$, where $c = 1$, $R_2 = 0.1$, $r = 0.973$, and $d_u = 6$.

Figure 9.4 shows the comparison results for the learning of the three algorithms. Figure 9.5 gives the illustration of the rule (neuron) evolution for the three algorithms during learning. Figure 9.6 shows the comparison results for the testing of the three algorithms. Table 9.1 shows the RMSE comparison results for the algorithms using (9.24).

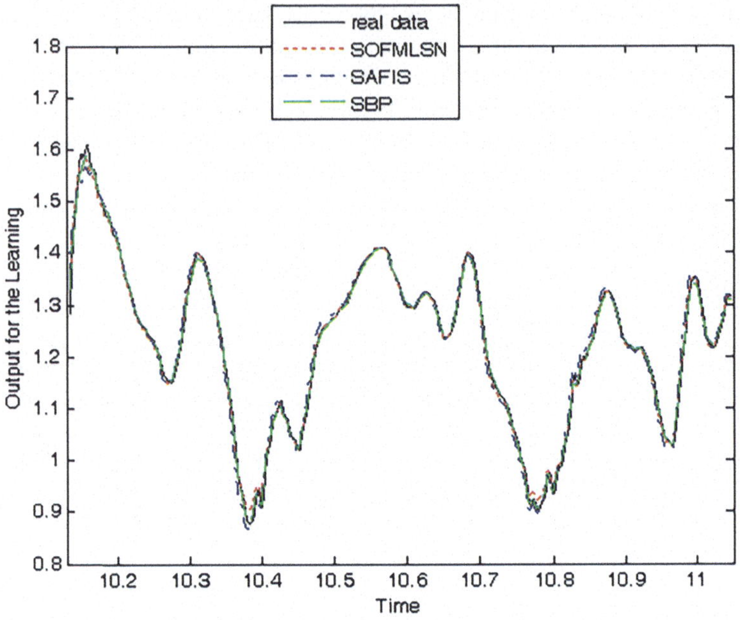

Fig. 9.4 Learning for Example 1

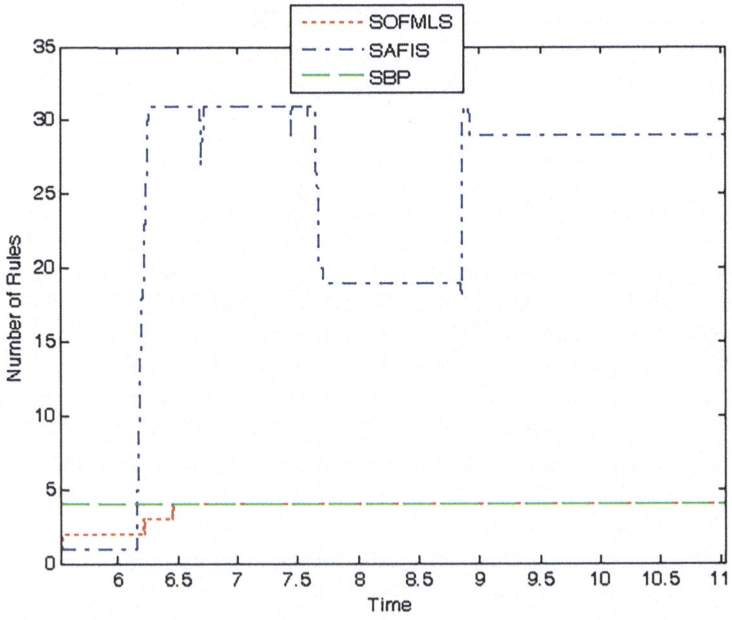

Fig. 9.5 Rule (neuron) evolution for Example 1

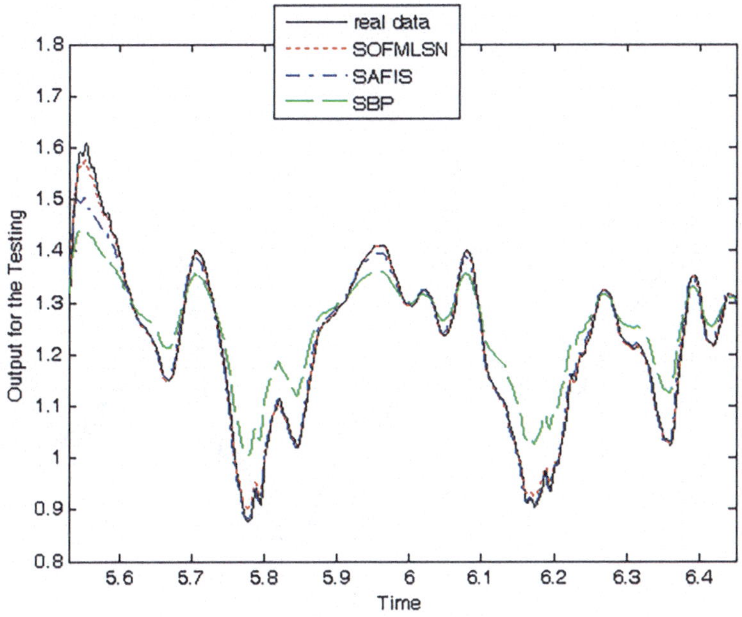

Fig. 9.6 Testing for Example 1

Table 9.1 Results for Example 1

Methods	Rules(Neurons)	Learning RMSE	Testing RMSE
SBP	4	0.0121	0.0233
SAFIS	29	0.0224	0.0077
SOFMLS	4	0.0118	0.0041

From Figs. 9.4, 9.5, and 9.6 and Table 9.1, it can be seen that the SOFMLS presents the smallest learning and testing RMSE, the SAFIS presents the biggest learning RMSE, the SBP presents the biggest testing RMSE, the SOFMLS and SBP give the smallest number of neurons, and the SAFIS gives the biggest number of neurons.

4.2 *Example 2*

Consider real data of eye signals of the up behavior [38] where 3572 pairs ($u(k)$, $y(k)$) of 3.572 s are used for the learning and 1192 pairs ($u(k)$, $y(k)$) for 1.192 s are used for the testing. The up signals are used in this chapter. The acquisition system is applied with a 25-year-old healthy man when his eyes are moving, and two electrodes are used to find the signals as described in the aforementioned section.

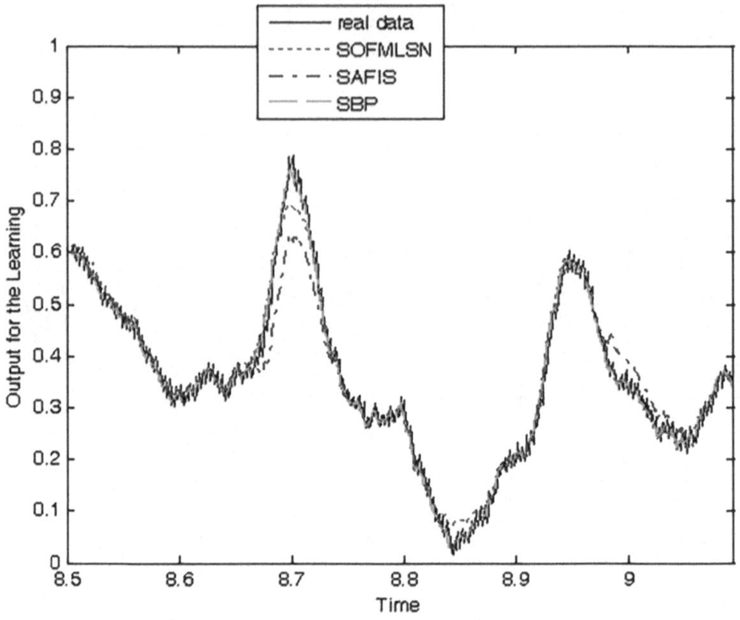

Fig. 9.7 Learning for Example 2

The inputs of all the intelligent systems are $y(k)$, $y(k+1)$, $y(k+2)$, $y(k+3)$, and the output of the intelligent systems is $y(k+4)$.

Considering Remark 9.3, the parameters for the SAFIS [16] are $N_x = 4$, $\gamma = 0.986$, $\varepsilon_{\max} = 0.1$, $n = 2$, $\varepsilon_{\min} = 0.01$, $e_g = 0.01$, $e_p = 0.001$. Considering Remark 9.7, the parameters of the SBP [22] are $N = 4$, $M = 3$, $\alpha_0 = 0.25$. Considering Remark 9.13, the parameters of the SOFMLS [28] are $N = 4$, $P_1 = cI \in \Re^{3x3}$, where $c = 1$, $R_2 = 0.1$, $r = 0.973$, and $d_u = 6$.

Figure 9.7 shows the comparison results for the learning of the three algorithms. Figure 9.8 gives the illustration of the rule (neuron) evolution for the three algorithms during learning. Figure 9.9 shows the comparison results for the testing of the three algorithms. Table 9.2 shows the RMSE comparison results for the algorithms using (9.24).

From Figs. 9.7, 9.8, and 9.9 and Table 9.2, it can be seen that the SOFMLS presents the smallest testing RMSE, the SBP presents the smallest learning RMSE, the SAFIS presents the biggest learning RMSE, the SBP presents the biggest testing RMSE, the SOFMLS gives the smallest number of neurons, and the SAFIS gives the biggest number of neurons.

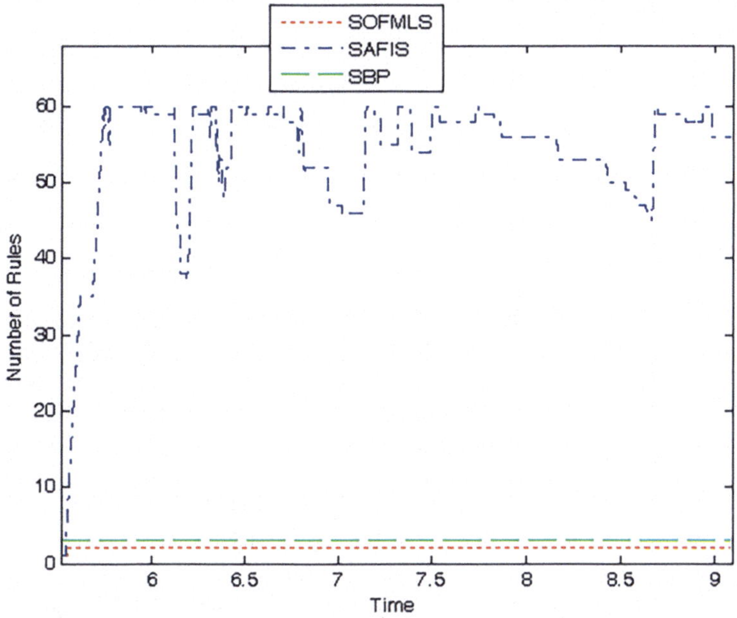

Fig. 9.8 Rule (neuron) evolution for Example 2

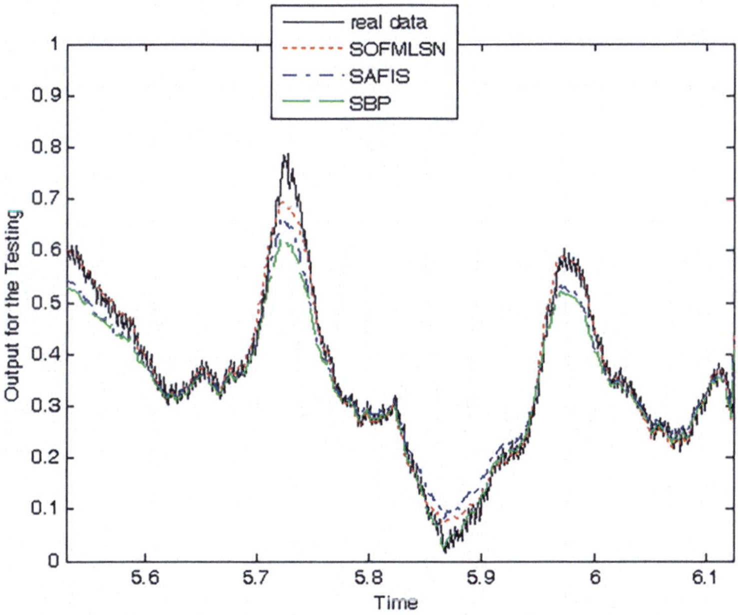

Fig. 9.9 Testing for Example 2

Table 9.2 Results for Example 2

Methods	Rules(Neurons)	Learning RMSE	Testing RMSE
SBP	3	0.0146	0.0285
SAFIS	56	0.0434	0.0259
SOFMLS	2	0.0190	0.0157

4.3 Example 3

Consider real data of brain signals [33] where 5528 pairs $(u(k), y(k))$ of 5.528 s are used for the learning and 1844 pairs $(u(k), y(k))$ for 1.844 s are used for the testing. The alpha signal is obtained in this study because it has more probabilities to be found. The acquisition system is applied with a 28-year-old healthy man when his eyes are closed. The inputs of all the intelligent systems are $y(k)$, $y(k+1)$, $y(k+2)$, $y(k+3)$, and the output of the intelligent systems is $y(k+4)$.

Considering Remark 9.3, the parameters for the SAFIS algorithm [16] are $N_x = 4$, $\gamma = 0.99$, $\varepsilon_{\max} = 1$, $n = 2$, $\varepsilon_{\min} = 0.1$, $e_g = 0.01$, $e_p = 0.001$. Considering Remark 9.7, the parameters of the SBP algorithm of [22] are $N = 4$, $M = 3$, $\alpha_0 = 0.5$. Considering Remark 9.13, the parameters of the SOFMLS algorithm [28] are $N = 4$, $P_1 = cI \in \Re^{3x3}$, where $c = 1$, $R_2 = 0.05$, $r = 0.93$, and $d_u = 6$.

Figure 9.10 shows the comparison results for the learning of the three algorithms. Figure 9.11 gives the illustration of the rule (neuron) evolution for the three

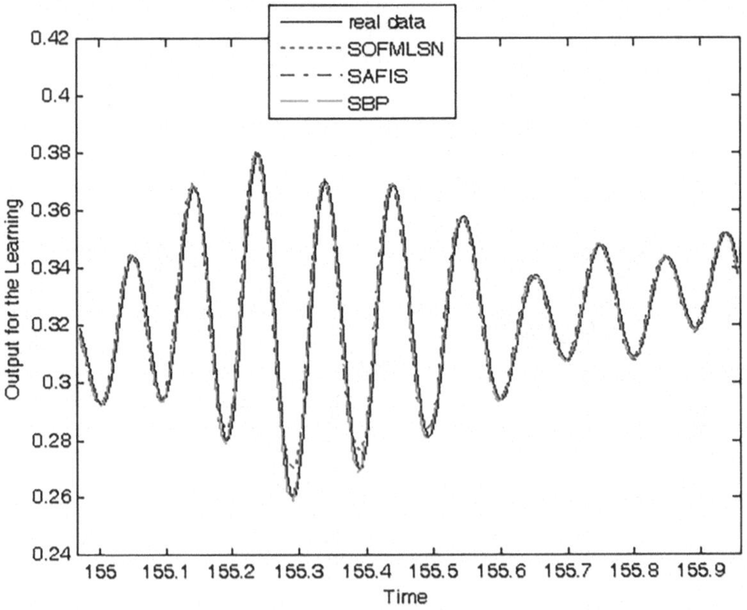

Fig. 9.10 Learning for Example 3

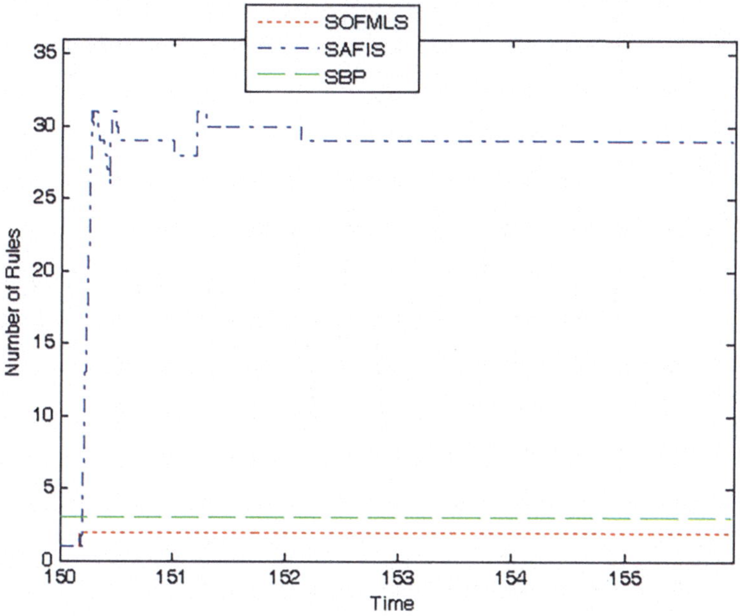

Fig. 9.11 Rule (neuron) evolution for Example 3

algorithms during learning. Figure 9.12 shows the comparison results for the testing of the three algorithms. Table 9.3 shows the RMSE comparison results for the algorithms using (9.24).

From Figs. 9.10, 9.11, and 9.12 and Table 9.3, it can be seen that the SOFMLS presents the smallest learning and testing RMSE, the SAFIS presents the biggest learning RMSE, the SBP presents the biggest testing RMSE, the SOFMLS gives the smallest number of neurons, and the SAFIS gives the biggest number of neurons.

4.4 *Example 4*

Consider real data of eye signals of the down behavior [38] where 3572 pairs ($u(k)$, $y(k)$) of 3.572 s are used for the learning and 1192 pairs ($u(k)$, $y(k)$) for 1.192 s are used for the testing. The up signals are used in this chapter. The acquisition system is applied with a 25-year-old healthy man when his eyes are moving, and two electrodes are used to find the signals as described in the aforementioned section. The inputs of all the intelligent systems are $y(k)$, $y(k+1)$, $y(k+2)$, $y(k+3)$, and the output of the intelligent systems is $y(k+4)$.

Considering Remark 9.3, the parameters for the SAFIS [16] are $N_x = 4$, $\gamma = 0.99$, $\varepsilon_{\max} = 0.1$, $n = 2$, $\varepsilon_{\min} = 0.01$, $e_g = 0.01$, $e_p = 0.001$. Considering Remark 9.7, the parameters of the SBP [22] are $N = 4$, $M = 3$,

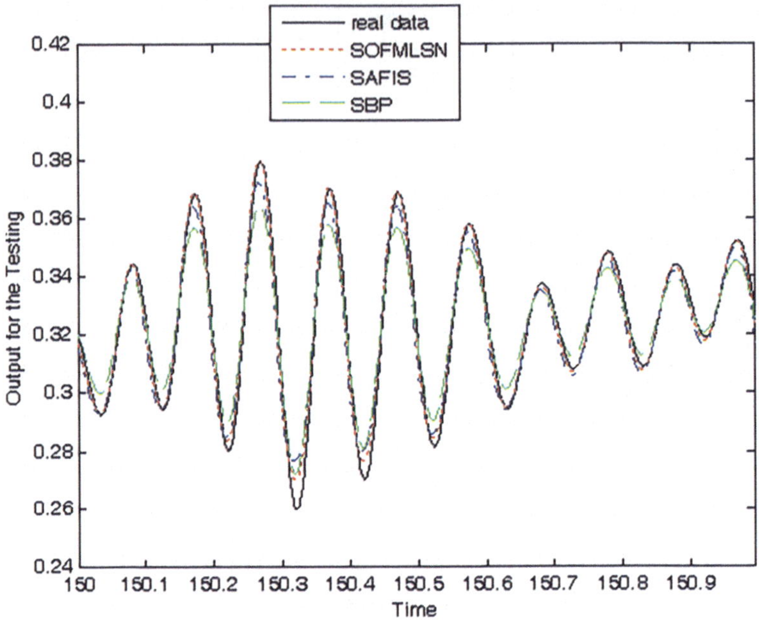

Fig. 9.12 Testing for Example 3

Table 9.3 Results for Example 3

Methods	Rules(Neurons)	Learning RMSE	Testing RMSE
SBP	3	0.0079	0.3443
SAFIS	29	0.0256	0.0077
SOFMLS	2	0.0067	0.0043

$\alpha_0 = 0.5$. Considering Remark 9.13, the parameters of the SOFMLS [28] are $N = 4$, $P_1 = cI \in \Re^{3x3}$, where $c = 1$, $R_2 = 0.05$, $r = 0.96$, and $d_u = 6$.

Figure 9.13 shows the comparison results for the learning of the three algorithms. Figure 9.14 gives the illustration of the rule (neuron) evolution for the three algorithms during learning. Figure 9.15 shows the comparison results for the testing of the three algorithms. Table 9.4 shows the RMSE comparison results for the algorithms using (9.24).

From Figs. 9.13, 9.14, and 9.15 and Table 9.4, it can be seen that the SOFMLS presents the smallest learning and testing RMSE, the SAFIS presents the biggest learning RMSE, the SBP presents the biggest testing RMSE, the SOFMLS and SBP give the smallest number of neurons, and the SAFIS gives the biggest number of neurons.

Remark 9.14 In the simulations, selecting different parameters for the algorithms of each example, the results present small variations.

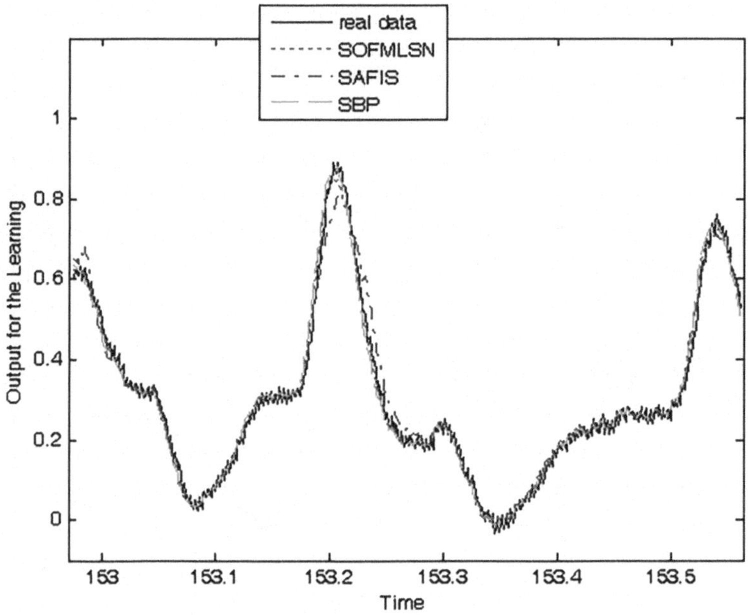

Fig. 9.13 Learning for Example 4

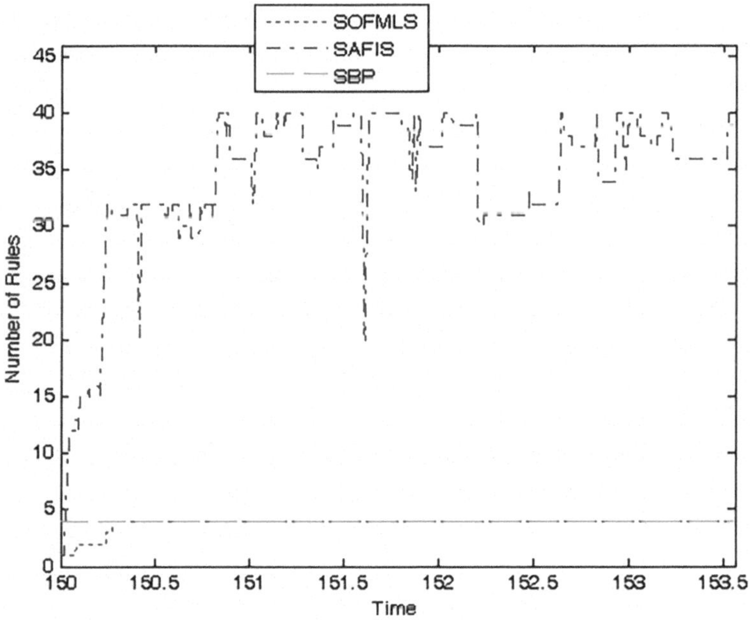

Fig. 9.14 Rule (neuron) evolution for Example 4

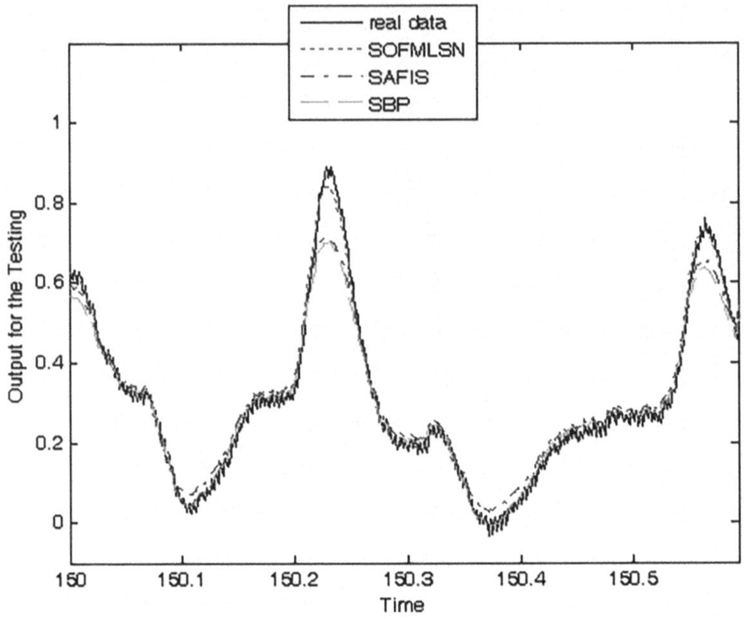

Fig. 9.15 Testing for Example 4

Table 9.4 Results for Example 4

Methods	Rules(Neurons)	Learning RMSE	Testing RMSE
SBP	4	0.0180	0.0305
SAFIS	40	0.0373	0.0297
SOFMLS	4	0.0146	0.0128

Remark 9.15 The SAFIS algorithm is applied in two synthetic examples and in the Makey-Glass time-series prediction problem [16]. The SBP algorithm is applied in a synthetic example and in the prediction of the loads distribution in a warehouse [22]. The SOFMLS algorithm is applied in two synthetic examples and in the Box-Jenkins furnace. This study is novel because it shows that the three algorithms can be used for the modeling of other different kind of systems which are the real brain and eye signals.

Remark 9.16 There is not a winner algorithm because the assumed tuning parameters for each method play their important role.

5 Concluding Remarks

This chapter successfully demonstrated the development of the SAFIS, SBP, and SOFMLS algorithms for the modeling of brain and eye signals. The simulation showed that the three algorithms can be used satisfactorily for the learning and testing of the real brain and eye signals. The learning could be applied for the control or prediction designs, and the testing could be applied for the classification, diagnosis, or prediction designs. The three methods can be used for the modeling of continuous and soft nonlinear systems or for the modeling of any of the conventional body signals. The three techniques are similar in that some parameters need to be decided in advance according to the problems considered, and other parameters are updated through the time. As a future work, some new evolving and stable intelligent algorithms will be designed.

References

1. P. Angelov, D. Filev, N. Kasabov, *Evolving Intelligent Systems, Methodology and Applications* (Wiley, New York, 2010)
2. M. Sayed-Mouchaweh, E. Lughofer, *Learning in Non-Stationary Environments: Methods and Applications* (Springer, New York, 2012)
3. D. Leite, P. Costa, F. Gomide, Interval approach for evolving granular system modeling, in *Learning in Non-Stationary Environments: Methods and Applications* (Springer, New York, 2012), pp. 273–304
4. E. Garcia-Cuesta, J.A. Iglesias, User modeling: through statistical analysis and subspace learning. Expert Syst. Appl. **39**, 5243–5250 (2012)
5. C.F. Juang, T.C. Chen, W.Y. Cheng, Speedup of implementing fuzzy neural networks with high dimensional inputs through parallel processing on graphic processing units. IEEE Trans. Fuzzy Syst. **19**(4), 717–728 (2011)
6. D. Leite, R. Ballini, P. Costa, F. Gomide, Evolving fuzzy granular modeling from nonstationary fuzzy data streams. Evol. Syst. **3**(2), 65–79 (2012)
7. A. Lemos, W. Caminhas, F. Gomide, Multivariable Gaussian evolving fuzzy modeling system. IEEE Trans. Fuzzy Syst. **19**(1), 91–104 (2011)
8. A. Lemos, W. Caminhas, F. Gomide, Fuzzy evolving linear regression trees. Evol. Syst. **2**(1), 1–14 (2011)
9. E. Lughofer, FLEXFIS: a robust incremental learning approach for evolving Takagi-Sugeno fuzzy models. IEEE Trans. Fuzzy Syst. **16**(6), 1393–1410 (2008)
10. E. Lughofer, P. Angelov, Handling drifts and shifts in online data streams with evolving fuzzy systems. Appl. Soft Comput. **11**(2), 2057–2068 (2011)
11. E. Lughofer, J.L. Bouchot, A. Shaker, On-line elimination of local redundancies in evolving fuzzy systems. Evol. Syst. **2**, 165–187 (2011)
12. E. Lughofer, Single pass active learning with conflict and ignorance. Evol. Syst. **3**, 251–271 (2012)
13. E. Lughofer, A dynamic split-and-merge approach for evolving cluster models. Evol. Syst. **3**, 135–151 (2012)
14. L. Maciel, A. Lemos, F. Gomide, R. Ballini, Evolving fuzzy systems for pricing fixed income options. Evol. Syst. **3**, 5–18 (2012)

15. F.J. Ordoñez, J.A. Iglesias, P. de Toledo, A. Ledezma, Online activity recognition using evolving classifiers. Expert Syst. Appl. **40**(4), 1248–1255 (2013)
16. H.J. Rong, N. Sundararajan, G.B. Huang, P. Saratchandran, Sequential adaptive fuzzy inference system (SAFIS) for nonlinear system identification and prediction. Fuzzy Sets Syst. **157**(9), 1260–1275 (2006)
17. H.J. Rong, N. Sundararajan, G.-B. Huang, G.-S. Zhao, Extended sequential adaptive fuzzy inference system for classification problems. Evol. Syst. **2**(2), 71–82 (2011)
18. H.J. Rong, S. Han, G.S. Zhao, Adaptive fuzzy control of aircraft wing-rock motion. Appl. Soft Comput. **14**(Part B), 181–193 (2014)
19. C.K. Ahn, An error passivation approach to filtering for switched neural networks with noise disturbance. Neural Comput. Appl. **21**(5), 853–861 (2012)
20. C.K. Ahn, M.T. Lim, Model predictive stabilizer for T-S fuzzy recurrent multilayer neural network models with general terminal weighting matrix. Neural Comput. Appl. **23**, 271–277 (2013)
21. X. Ren, X. Lv, Identification of extended Hammerstein systems using dynamic self-optimizing neural networks. IEEE Trans. Neural Netw. **22**(8), 1169–1179 (2011)
22. J.J. Rubio, P. Angelov, J. Pacheco, An uniformly stable backpropagation algorithm to train a feedforward neural network. IEEE Trans. Neural Netw. **22**(3), 356–366 (2011)
23. J.J. Rubio, D.M. Vazquez, J. Pacheco, Backpropagation to train an evolving radial basis function neural network. Evol. Syst. **1**(3), 173–180 (2010)
24. X. Wang, Y. Huang, Convergence study in an extended Kalman filter-based training of recurrent neural networks. IEEE Trans. Neural Netw. **22**(4), 588–600 (2011)
25. W. Yu, J.J. Rubio, Recurrent neural network training with stable bounding ellipsoid algorithm. IEEE Trans. Neural Netw. **20**(6), 983–991 (2009)
26. W. Zhang, W. Wu, M. Yao, Boundedness and convergence of bath backpropagation algorithm with penalty with feedforward neural networks. Neurocomputing **89**(15), 141–146 (2012)
27. E. Lughofer, *Evolving Fuzzy Systems, Methodologies, Advanced Concepts and Applications* (Springer, Berlin, 2011)
28. J.J. Rubio, SOFMLS: online Self-organizing fuzzy modified least square network. IEEE Trans. Fuzzy Syst. **17**(6), 1296–1309 (2009)
29. J.J. Rubio, J.H. Perez Cruz, Evolving intelligent system for the modelling of nonlinear systems with dead-zone input. Appl. Soft Comput. **14**(Part B), 289–304 (2014)
30. C. Martinez, B. Rojas, Técnicas de Electroencefalografía, 2da edición, Secretaria de Educación Pública, Comunicaciones Científicas Mexicanas S.A. de C.V., México (1998). ISBN: 968–7858-12-5
31. J.B. Webster, *Medical Instrumentation, Application and Design*, 4th edn. (Wiley, United States of America, 2010)
32. C. Ramirez, M. Hernandez, Procesamiento en tiempo real de variables Fisiológicas, in Universidad Nacional de Experimental de Táchira, Decanato de Investigación, Grupo de Biomédica
33. J.J. Rubio, D.M. Vazquez, D. Mujica-Vargas, Acquisition system and approximation of brain signals. IET Sci. Meas. Technol. **7**(4), 232–239 (2013)
34. F. Gibbs, E. Gibbs, Atlas of electroencephalography, Changes whit age, asleep., Addison Wesley, Massachusetts, 1, 82–89, (1950).
35. D. Klass, D. Daly, *Current Practice of Clinical Electroencephalography, Chap 5* (Raven Press, New York, 1975), pp. 69–109
36. S. De Castro, J. Perez, *Manual de patología general, 6ta Edición* (Masson Elsevier, Barcelona, 2006). ISBN: 978 84 458 1540 3
37. J. Hori, K. Sakano, Y. Saitoh, Development of a communication support device controlled by eye movements and voluntary eye blink. IEICE Trans. Inf. Syst. **E89D**(6), 1790–1797 (2006)
38. J.J. Rubio, F. Ortiz, C.R. Mariaca, J.C. Tovar, A method for online pattern recognition for abnormal eye movements. Neural Comput. Appl. **22**(3–4), 597–605 (2013)

39. K. Yamagishi, J. Hori, M. Miyakama, Development of EOG-based communication system controlled by eight-directional eye movements, in *Proceedings of the 28th IEEE EMBS Annual International Conference* (2006), pp. 2574–2577
40. D.M. Vazquez, J.J. Rubio, J. Pacheco, A characterization framework for epileptic signals. IET Image Process. **6**(9), 1227–1235 (2012)

Chapter 10
MSAFIS: An Evolving Fuzzy Inference System

1 Introduction

The recent years have witnessed the emergence of an important topic related to process learning which is learning from big data (LBD). LBD is concerned with the development and application of learning algorithms for very large, possibly complex, datasets that cannot be accommodated in the main memory. To cope with this requirement, different techniques and technologies have been proposed:

1. Parallel and distributed computing (e.g., Hadoop): Data are split into portions and sent to parallel machines to be processed and learned from.
2. Online learning, known also as sequential learning, one-pass learning, real-time learning, evolving systems, etc.: The learning algorithms learn sequentially, either batch-based or point-based, potentially using one single machine.

Although these techniques are not new from a pure scientific point of view, the deluge of data available everywhere has given a refreshing and renewable interest to them. In this chapter we will focus on online learning.

Online learning faces the challenge of accurately estimating models using incoming data whose statistical characteristics are not known a priori. In nonstationary environments, the challenge becomes even more important, since the model's behavior may need to change drastically over time [1]. Online learning aims at ensuring continuous adaptation of the model being fitted to the data. When learning, ideally only the model should be stored in memory. For instance, in rule-based systems (RBSs), only rules should be memorized. The model is then adjusted in future learning steps. In the case of RBS, as new data arrive, new rules may be created, and existing ones may be modified or removed allowing the overall model to evolve over time [2] and [3]. In [4], online fuzzy models are discussed.

In general evolving systems are online learning algorithms whose structure and parameters are very flexible in order to adapt to ever-changing environments [5–10]. Online processing of data with a particular focus on the design issues of online

J. de J. Rubio, *Stability Analysis of Neural Networks and Evolving Intelligent Systems*, https://doi.org/10.1007/978-3-031-87282-2_10

evolving systems is considered in [11]. In [2], online self-learning fuzzy classifier, called GT2FC standing for "Growing Type-2 Fuzzy Classifier," is presented. The proposed approach shows how type-2 fuzzy rules can be learned online in an evolving way from data streams. GT2FC was applied in the context of smart homes. In [12], the authors explore the application of interactive and online learning of user profiles in the context of information filtering using evolutionary algorithms. In [13], an evolving algorithm for learning computer user behavior is introduced.

Evolving systems have been very popular, for instance, in [14], a learning approach to train uninorm-based hybrid neural networks is mentioned. The use of evolving classifiers for activity recognition is described in [15] and [16]. In [17, 18], and [19], novel efficient techniques of evolving intelligent systems are discussed. A dynamic pattern recognition method is introduced in [20]. In [21], an approach for classifying huge amounts of different news articles is designed. An evolving method that is able to keep track of computer users is proposed in [13]. In [22], a new approach called evolving principal component clustering is addressed. A new clustering method is suggested in [23]. In [24] and [25], novel evolving fuzzy-rule-based classifiers are addressed. An evolving neural fuzzy modeling approach is constructed in [26]. In [27], a novel approach in fault diagnosis is studied.

Stable systems are characterized by the boundedness criterion, i.e., if bounded algorithm inputs are employed, then the outputs and parameters exponentially decay to a small and bounded zone. In [28], the author uses an induced $L\infty$ approach to create a new filter with a finite impulse response structure for state-space models with external disturbances. The model predictive stabilization problem for Takagi-Sugeno fuzzy multilayer neural networks with general terminal weighting matrix is investigated in [29]. In [30], an error passivation approach is used to derive a new passive and exponential filter for switched Hopfield neural networks with time delay and noise disturbance. Two robust intelligent controllers for nonlinear systems with dead-zone are addressed in [31] and [32]. In [33] and [34], two stable controllers are introduced.

However, most of these algorithms operate offline and are not designed to handle big data. The present chapter presents the combination of two algorithms: the sequential adaptive fuzzy inference system (SAFIS) [35] which is an evolving algorithm and the stable gradient descent algorithm (SGD) [3] which is a stable algorithm. Such a combination, called the MSAFIS, aims to devise an efficient evolving algorithm that can cope with data streams as a case of big data. MSAFIS exploits the SGD algorithm to update parameters, while SAFIS relies on the Kalman filter. SGD has the advantage that it outperforms Kalman filter [3].

The chapter is organized as follows. In Sect. 2, the SAFIS, SGD, and MSAFIS algorithms are detailed. In Sect. 3, the brain encephalography (EEG) and the eye electrooculogram (EOG) signals are described. Using an EEG and an EOG dataset, SAFIS, SGD, and MSAFIS are evaluated and compared in Sect. 4. Section 5 concludes the chapter and suggests future research directions.

2 Presentation of the Algorithms

In this section the three algorithms SAFIS, SGD, and MSAFIS are described. Furthermore, the differences of the three algorithms are explained.

2.1 SAFIS Algorithm

The sequential adaptive fuzzy inference system (SAFIS) is developed based on the functional equivalence between a radial basis function network and a fuzzy inference system (FIS) resulting in a neuro fuzzy system. In SAFIS, the concept of "Influence" of a fuzzy rule is introduced, and using this the fuzzy rules are added or removed based on the input data received so far. If the input data do not warrant adding of fuzzy rules, then only the parameters of the "closest" (in a Euclidean sense) rule are updated using an extended Kalman filter (EKF) scheme.

The SAFIS algorithm is summarized as below [35]:

For each observation $(z(k), y(k))$, where $z(k) \in \Re^N$, $y(k) \in \Re$, and $k = 1, 2, \ldots,$ do:

(1) Compute the overall system output:

$$\widehat{y}(k) = \frac{\sum_{j=1}^{M} o_j(k) R_j(z_i(k))}{\sum_{j=1}^{M} R_j(z_i(k))}, \tag{10.1}$$

where

$$R_j(z_i(k)) = \exp\left(-\frac{1}{\delta_j^2(k)} \left\| z_i(k) - m_j(k) \right\|^2 \right),$$

and M is the number of fuzzy rules, $R_j(z_i(k))$ is the firing strength of the jth rule, and $o_j(k)$ is the weight of the normalized rule. Note that each rule is represented as a radial basis function described by its center $m_j(k)$ and its spread $\delta_j(k)$.

(2) Calculate the parameters required in the growth criterion:

$$\epsilon(k) = \max\left\{ \epsilon_{\max} \tau^k, \epsilon_{\min} \right\}, 0 < \tau < 1, \tag{10.2}$$

where $\epsilon_{\max}$ and $\epsilon_{\min}$ are the threshold largest and smallest distances admitted between the inputs and the corresponding nearest center of rules. The parameter τ $(0 < \tau < 1)$ indicates the decay constant. The error of the kth input is given as follows:

$$\widetilde{y}(k) = y(k) - \widehat{y}(k). \tag{10.3}$$

where $y(k)$ and $\widehat{y}(k)$ are the output and the estimated output, respectively.

(3) Apply the criterion for adding rules if the following two conditions are satisfied:

If

$$\left\| z_i(k) - m_j(k) \right\| > \epsilon(k), \tag{10.4}$$

and

$$Y_{\text{inf}}(M+1) = |\widetilde{y}(k)| \frac{\left(1.8K \left\| z_i(k) - m_j(k) \right\|\right)^N}{\sum_{j=1}^{M+1} (1.8\delta_j(k))^N} > y_g, \tag{10.5}$$

where y_g is the growing threshold. A new rule $M+1$ is added if y_g is exceeded. The new rule $M+1$ is given as follows:

$$\begin{aligned} o_{M+1}(k) &= \widetilde{y}(k), \\ m_{M+1}(k) &= z_i(k), \\ \delta_{M+1}(k) &= K \left\| z_i(k) - m_{M+1}(k) \right\|. \end{aligned} \tag{10.6}$$

If no rule is added, the nearest rule jm is obtained as follows:

$$\min_j R_j(z(k)) \Longrightarrow jm = j, \tag{10.7}$$

and adjust the system parameters $o_j(k)$, $m_j(k)$, $\delta_j(k)$ for the nearest rule only by using the extended Kalman filter (EKF) method:

$$\begin{aligned} \varphi(k) &= \varphi(k-1) + P_{k-1}b(k-1)\left[a + b^T(k-1)P_{k-1}b(k-1)\right]^{-1}\widetilde{y}(k), \\ P_k &= P_{k-1} - P_{k-1}b(k-1)\left[p + b^T(k-1)P_{k-1}b(k-1)\right]^{-1} \\ &\quad b^T(k-1)P_{k-1} + qI, \end{aligned} \tag{10.8}$$

where $\varphi(k) = [\varphi_1(k) \cdots \varphi_3(k)]^T = [m_{jm}(k),\ o_{jm}(k),\ \delta_{jm}(k)]^T$, $P_1 = qI$, q and p are parameters selected by the designer, $0 < q < 1$, $0 <$

$p < 1$, $b(k)=[b_1(k), b_2(k), b_3(k)]^T$, $b_1(k)=\frac{2[o_{jm}(k)-\widehat{y}(k)]R_{jm}(z_i(k))[z_i(k)-m_{jm}(k)]}{\left[\sum_{j=1}^{M} R_j(z_i(k))\right]\delta_{jm}^2(k)}$,

$b_2(k)=\frac{2[o_{jm}(k)-\widehat{y}(k)]R_{jm}(z_i(k))\|z_i(k)-m_{jm}(k)\|^2}{\left[\sum_{j=1}^{M} R_j(z_i(k))\right]\delta_{jm}^3(k)}$, $b_3(k)=\frac{R_{jm}(z_i(k))}{\left[\sum_{j=1}^{M} R_j(z_i(k))\right]}$, and I is the identity matrix.

(4) If the following criterion is satisfied:

$$Y_{\inf}(jm) = |o_{jm}(k)| \frac{(1.8\delta_{jm}(k))^N}{\sum_{j=1}^{M}(1.8\delta_j(k))^N} < y_p, \tag{10.9}$$

then, remove the jm rule and reduce the dimensionality of EKF. Note that y_p is the pruning threshold.

Remark 10.1 The significance of a rule proposed in growing and pruning radial basis function (GAP-RBF) neural network is defined based on the average contribution of an individual rule to the output of the RBF network. Under this definition, one may need to estimate the input distribution range $S(z) = \frac{|o_{jm}(k)|}{\sum_{j=1}^{M}(1.8\delta_j(k))^N}$. However, the influence of a rule introduced in this chapter is different from the significance of a rule proposed in GAP-RBF. In fact, the influence of a rule is defined as the relevant significance of the rule compared to summation of significance of all the existing RBF rules. As seen from Eq. (10.7), with the introduction of influence one need not estimate the input distribution range, and the implementation has been simplified.

Remark 10.2 In parameter modification, SAFIS utilizes a winner rule strategy similar to the work done by Huang et al. [36]. The key idea of the winner rule strategy is that only the parameters related to the selected winner rule are updated by the EKF algorithm in every step. The "winner rule" is defined as the rule that is closest (in the Euclidean distance sense) to the current input data. As a result, SAFIS is computationally efficient.

Remark 10.3 In SAFIS, some parameters need to be decided in advance according to the problems considered. They include the distance thresholds ($\epsilon_{\max}$, $\epsilon_{\min}$, τ), the overlap factor K for determining the width of the newly added rule, the growing threshold (y_g) for a new rule, and the pruning threshold (y_p) for removing an insignificant rule. A general selection procedure for the predefined parameters is given as follows: max is set to around the upper bound of input variables, $\epsilon_{\min}$ is set to around 10% of $\epsilon_{\max}$, and τ is set to around 0.99. y_p is set to around 10% of y_g.

$\epsilon_{\max}$ is observed in the range [1.0, 10.0]. The overlap factor K is utilized to initialize the width of the newly added rule and chosen according to different problems; it is observed in the range [1.0, 2.0]. The growing threshold y_g is chosen according to the system performance; it is observed in the range [0.001, 0.05]. The smaller the y_g, the better the system performance, but the resulting system structure is more complex.

2.2 SGD Algorithm

The stable gradient descent (SGD) algorithm is developed with a new time-varying rate to guarantee its uniformly stability for online identification and its identification error converges to a small zone bounded by the uncertainty. The weights' error is bounded by the initial weights' error, i.e., hence the overfitting is avoided. The SGD algorithm is as follows [3]:

(1) Compute the output of the nonlinear system $y(k)$ with Eq. (10.10). Note that the nonlinear system may have the structure represented by Eq. (10.10), and the parameter N is selected according to this nonlinear system.

$$y(k) = f\left[z(k)\right], \tag{10.10}$$

where $z(k) = [z_1(k)\ldots, z_i(k), \ldots, z_N(k)]^T = [y(k-1), \ldots, y(k-n), u(k-1), \ldots, u(k-m)]^T \in \Re^{N\times 1}$ $(N = n+m)$ is the input vector, $u(k-1) \in \Re$ is the input of the plant, $y(k) \in \Re$ is the output of the plant, and f is an unknown nonlinear function, $f \in C^{\infty}$.

(2) Select the following parameters: $o(1)$ and $w(1)$ as random numbers between 0 and 1, M as an integer number, and α_0 as a positive value smaller or equal to 1; obtain the output $\widehat{y}(1)$ using Eq. (10.11).

$$\begin{aligned} \widehat{y}(k) &= \sum_{j=1}^{M} o_j(k)\beta_j(k), \\ \beta_j(k) &= \tanh\Big(\sum_{i=1}^{N} w_{ij}(k)z_i(k)\Big). \end{aligned} \tag{10.11}$$

(3) For each iteration k, obtain the output $\widehat{y}(k)$ with Eq. (10.11), also obtain the identification error $\widetilde{y}(k)$ with Eq. (10.12):

$$\widetilde{y}(k) = \widehat{y}(k) - y(k), \tag{10.12}$$

and update the parameters $o_j(k)$ and $w_{ij}(k)$ using Eq. (10.13):

$$\begin{aligned} o_j(k) &= o_j(k-1) - \alpha(k-1)\beta_j(k-1)\widetilde{y}(k-1), \\ w_{ij}(k) &= w_{ij}(k-1) - \alpha(k-1)\gamma_{ij}(k-1)\widetilde{y}(k-1), \end{aligned} \tag{10.13}$$

where the new time varying rate $\alpha(k)$ is

$$\alpha(k-1) = \frac{\alpha_0}{2\left(\frac{1}{2} + \sum_{j=1}^{M}\beta_j^2(k-1) + \sum_{j=1}^{M}\sum_{i=1}^{N}\gamma_{ij}^2(k-1)\right)},$$

where $i = 1, \ldots, N$, $j = 1, \ldots, M$, $\gamma_{ij}(k-1) = o_j(k)\text{sech}^2(\sum_{i=1}^{N} w_{ij}(k-1)z_i(k-1))z_i(k-1) \in \Re$.

Remark 10.4 There are two conditions for applying this algorithm for nonlinear systems: The first one is that the nonlinear system may have the form described by (10.10), and the second one is that the uncertainty $\mu(k) = y(k) - \sum_{j=1}^{M} o_j^* \beta_j^*$ may be bounded, $\beta_j^* = \tanh(\sum_{i=1}^{N} w_{ij}^* z_i(k))$, and o_j^* and w_{ij}^* are unknown weights such that the uncertainty $\mu(k)$ is minimized.

Remark 10.5 The value of the parameter used for the stability of the algorithm $\overline{\mu}$ is unimportant, because this parameter is not used in the algorithm. The bound of $\mu(k)$ is needed to guarantee the stability of the algorithm, but it is not used in the SGD algorithm (10.11), (10.12), (10.13).

Remark 10.6 The proposed SGD has one hidden layer. It was reported in the literature that a feedforward neural network with one hidden layer is enough to approximate any nonlinear system.

Remark 10.7 Note that the behavior of the algorithm could be improved or deteriorated by changing the values of M or α_0.

2.3 MSAFIS

The MSAFIS is the SAFIS algorithm with the modification of Eqs. (10.3) and (10.8) by Eqs. (10.12) and (10.13) and using the parameters of the SAFIS algorithm $m_j(k)$, $\delta_j(k)$, $o_j(k)$ instead of the parameters of the SGD algorithm $w_{ij}(k)$, $o_j(k)$. The MSAFIS algorithm is summarized as follows.

For each observation $(z(k), y(k))$, where $z(k) \in \Re^N$, $y(k) \in \Re$, and $k = 1, 2, \ldots,$ do:

(1) Compute the overall system output:

$$\widehat{y}(k) = \frac{\sum_{j=1}^{M} o_j(k) R_j(z_i(k))}{\sum_{j=1}^{M} R_j(z_i(k))}, \tag{10.14}$$

where

$$R_j(z_i(k)) = \exp\left(-\frac{1}{\delta_j^2(k)} \left\| z_i(k) - m_j(k) \right\|^2\right),$$

and M is the number of fuzzy rules, $R_j(z_i(k))$ is the firing strength of the jth rule, and $o_j(k)$ is the weight of the normalized rule. Note that each rule is represented as a radial basis function described by its center $m_j(k)$ and its spread $\delta_j(k)$.

(2) Calculate the parameters required in the growth criterion:

$$\epsilon(k) = \max\left\{\epsilon_{\max}\tau^k, \epsilon_{\min}\right\}, 0 < \tau < 1, \tag{10.15}$$

where $\epsilon_{\max}$ and $\epsilon_{\min}$ are the threshold largest and smallest distances admitted between the inputs and corresponding nearest center of rules. The parameter τ $(0 < \tau < 1)$ indicates the decay constant. The error of the kth input is given as follows:

$$\widetilde{y}(k) = \widehat{y}(k) - y(k), \tag{10.16}$$

(3) Apply the criterion for adding rules if the following two conditions are satisfied:

If

$$\left\| z_i(k) - m_j(k) \right\| > \epsilon(k), \tag{10.17}$$

and

$$Y_{\inf}(M+1) = |\widetilde{y}(k)| \frac{\left(1.8K \left\| z_i(k) - m_j(k) \right\|\right)^N}{\sum_{j=1}^{M+1} (1.8\delta_j(k))^N} > y_g. \tag{10.18}$$

where y_g is the growing threshold. A new rule $M+1$ is added if y_g is exceeded.

The new rule $M+1$ is given as follows:

$$\begin{aligned} o_{M+1}(k) &= \widetilde{y}(k), \\ m_{M+1}(k) &= z_i(k), \\ \delta_{M+1}(k) &= K \left\| z_i(k) - m_{M+1}(k) \right\| . \end{aligned} \tag{10.19}$$

If no rule is added, the nearest rule jm is obtained as follows:

$$\min_j R_j(z(k)) \Longrightarrow jm = j, \tag{10.20}$$

and adjust the system parameters $o_j(k)$, $m_j(k)$, $\delta_j(k)$ for the nearest rule only by using the stable gradient descent algorithm:

$$\varphi(k) = \varphi(k-1) - \alpha(k-1)b(k-1)\widetilde{y}(k-1), \tag{10.21}$$

where $\varphi(k)=[\varphi_1(k),\ \varphi_2(k),\ \varphi_3(k)]^T=[m_{jm}(k),\ o_{jm}(k),\ \delta_{jm}(k)]^T$, $b(k)=[b_1(k),\ b_2(k),\ b_3(k)]^T$, $b_1(k)=\frac{2[o_{jm}(k)-\widehat{y}(k)]R_{jm}(z_i(k))[z_i(k)-m_{jm}(k)]}{\left[\sum_{j=1}^{M} R_j(z_i(k))\right]\delta_{jm}^2(k)}$, $b_2(k)=\frac{2[o_{jm}(k)-\widehat{y}(k)]R_{jm}(z_i(k))\|z_i(k)-m_{jm}(k)\|^2}{\left[\sum_{j=1}^{M} R_j(z_i(k))\right]\delta_{jm}^3(k)}$, $b_3(k)=\frac{R_{jm}(z_i(k))}{\left[\sum_{j=1}^{M} R_j(z_i(k))\right]}$, and the new time varying rate $\alpha(k-1)$ is

$$\alpha(k-1) = \frac{\alpha_0}{2\left(\frac{1}{2} + \sum_{l=1}^{3} b_l^2(k-1)\right)},$$

where α_0 is a parameter selected by the designer, $0 < \alpha_0 < 1$.

(4) If the following criterion is satisfied:

If

$$Y_{\inf}(jm) = \left|o_{jm}(k)\right| \frac{\left(1.8\delta_{jm}(k)\right)^N}{\sum_{j=1}^{M}(1.8\delta_j(k))^N} < y_p, \tag{10.22}$$

then, remove the jm rule, and reduce the dimensionality of SGD. Note that y_p is the pruning threshold.

Remark 10.8 In MSAFIS, some parameters need to be decided in advance according to the problems considered. They include the distance thresholds ($\epsilon_{\max}$, $\epsilon_{\min}$, τ), the overlap factor K for determining the width of the newly added rule, the growing threshold (y_g) for a new rule, and the pruning threshold (y_p) for removing an insignificant rule. A general selection procedure for the predefined parameters is given as follows: max is set to around the upper bound of input variables, ϵ_{min} is set to around 10% of ϵ_{max}, and τ is set to around 0.99. y_p is set to around 10% of y_g. ϵ_{max} is observed in the range [1.0, 10.0]. The overlap factor K is utilized to initialize the width of the newly added rule and chosen according to different problems; it is observed in the range [1.0, 2.0]. The growing threshold y_g is chosen according to the system performance; it is observed in the range [0.001, 0.05]. The smaller the y_g, the better the system performance, but the resulting system structure is more complex.

2.4 Comparison of the Three Algorithms

In this subsection, the comparison between the three algorithms is described.

Table 10.1 shows several aspects about the three algorithms.

Table 10.2 shows an overview of the modifications made to the SAFIS to evolve the new method, called MSAFIS.

Note that the SGD is not included in Table 10.2 because it is more different than the other two algorithms.

Table 10.1 Characteristics of the three algorithms

SAFIS	SGD	MSAFIS
If it is applied to systems which have important changes through the time, an acceptable result can be assured	If it is applied to systems which have important changes through the time, an acceptable result cannot be assured	If it is applied to systems which have important changes through the time, an acceptable result can be assured
If it is applied to unstable systems, an acceptable result cannot be assured	If it is applied to unstable systems, an acceptable result can be assured	If it is applied to unstable systems, an acceptable result can be assured
It can be applied in many systems as are the biology, mechatronic, mechanic, thermal, robotic, economic, etc.	It can be applied in many systems as are the biology, mechatronic, mechanic, thermal, robotic, economic, etc.	It can be applied in many systems as are the biology, mechatronic, mechanic, thermal, robotic, economic, etc.

Table 10.2 Differences between the SAFIS and MSAFIS

SAFIS	MSAFIS
Equation (10.3). The error is obtained by subtracting the estimated output to the output	Equation (10.16). The error is obtained by subtracting the output to the estimated output
Equation (10.8). The parameters are adjusted using the extended Kalman filter algorithm	Equation (10.21). The parameters are adjusted using the stable gradient descent algorithm

3 The Brain and Eye Signals

This section describes the characteristics of the brain and eye signals.

3.1 The EEG Signals

The difference of the potential in one membrane is obtained by the exchange between the ions (Na+,Cl−,K+) being in the same. The rules have a potential difference between the inside and outside which is called rest potential, and this potential represents constant changes because of the impulses given by the neighbor rules. This potential difference can be measured in the brain cortex using electrodes that convert the ion flow into electric flow. The characteristic of the encephalography signal (EEG) is of 5−300 μV in amplitude and of 0−150 Hz in frequency [37].

The EEG signals are waves similar to periodic, but the waves can change from one time to other, and they have some characteristics that allow the learning, as are the amplitude, the frequency, the morphology, the band, the rhythm, and the duration [37].

The following paragraphs show the characteristics which are considered for an adult in vigilance [37].

Alpha signal. It is the normal rhythm of the bottom and is the most stable and typical in the human. It is found in the frequencies of 8−12 Hz ± 1 Hz. The amplitude is between 20 and 60 μV. It can be seen generally in posterior regions with more amplitude in the occipital lobes. See Fig. 10.1. It is more evident when the patient is awake with closed eyes and in physical and mental rest, and it is stopped when the eyes are opened or with the mental activity.

Beta signal. It is found in the frequencies >13 Hz, in general between 14 and 35 Hz. The amplitude is usually low from 5 to 10 μV and is symmetric. See Fig. 10.1.

Theta signal. It has a frequency of 4−8 Hz, is of half of low voltage, and is found in the temporal regions. See Fig. 10.1.

Delta signal. It is found in the second and the third stages of the dream. It has a frequency of 0.5−3.5 Hz, and the amplitude is generally higher than 75 μV. See Fig. 10.1.

3.2 The EOG Signals

The electrooculograms (EOGs) are the signals obtained as a result of the eye movements of a patient, and these EOGs are detected using three electrodes: one electrode on the temple, one above, and other underneath of the eye. Usually, the detected signals are by direct current (DC) coupling to specify the direction of the

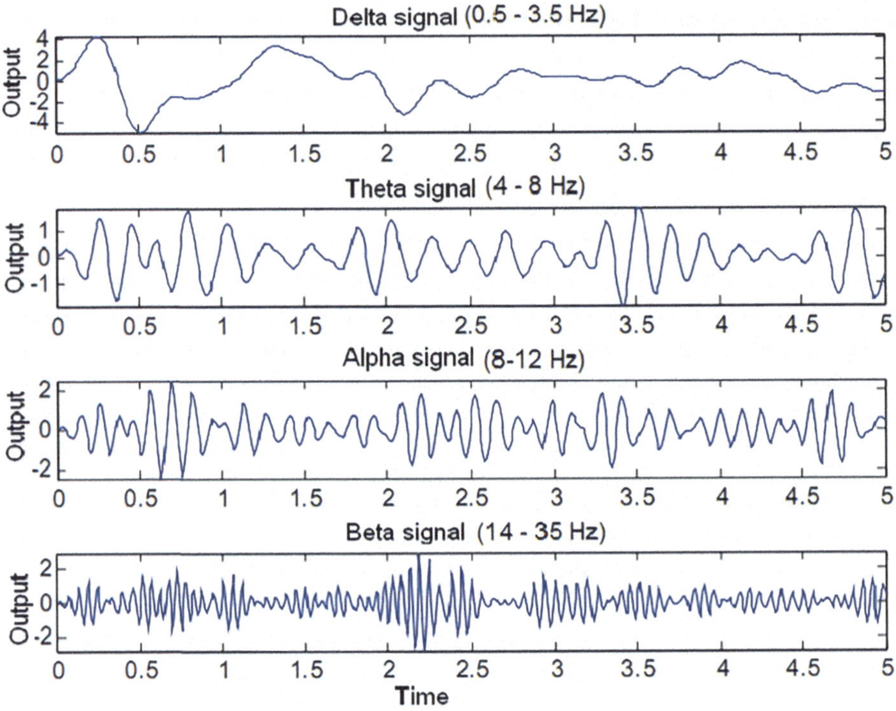

Fig. 10.1 EEG signals

gaze. In the experiments of this chapter, three electrodes are placed on the dominant side of the patient eye according to the optimum positions suggested by Rubio et al. [38].

Figure 10.2 shows the relationship between real eye movements (input) and the EOG signals (output) of the system. Denote the upper and lower thresholds of the vertical channel Ch.V as V1 and V2, respectively, and denote the upper and lower thresholds of the horizontal channel Ch.H as H1 and H2, respectively. When the EOG potential exceeds one of these thresholds, the output assumes ON, and when the EOG potential does not exceed one of these thresholds, the output assumes OFF. The process of transforming the EOG signals from the intention of the patient is as follows [38]:

1. Output Up is when it is obtained an Up behavior: First, Threshold V1 of the vertical channel becomes ON, while Threshold V2 is OFF, and second, Threshold V2 of the vertical channel becomes ON, while Threshold V1 becomes OFF. H1 and H2 of the horizontal channel remain OFF all the time.

2. Output Down is when it is obtained a Down behavior: First, Threshold V2 of the vertical channel becomes ON, while Threshold V1 is OFF, and second, Threshold V1 of the vertical channel becomes ON, while Threshold V2 becomes OFF. H1 and H2 of the horizontal channel remain OFF all the time.

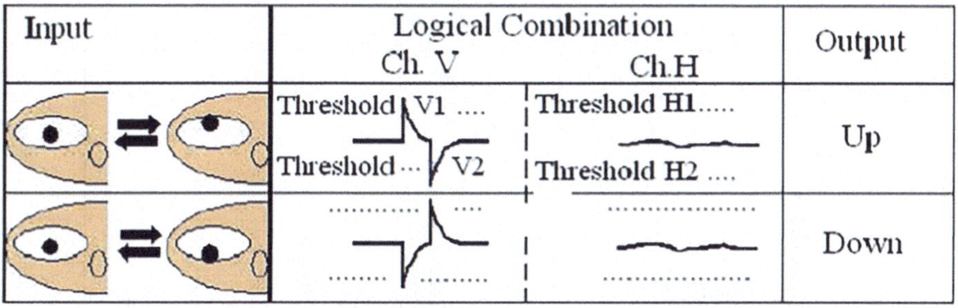

Fig. 10.2 EOG signals

4 Results

In this section, the three above detailed algorithms are applied for the learning of brain and eye signals with big data. The aforementioned signals could be applied for patient who cannot move their bodies; consequently, they could use their brains or their eyes to say what they want or need. The SAFIS of [35], SGD of [3], and MSAFIS are compared for the learning sequentially:

- Brain signals: experiment 1
- Eye signals: experiment 2

In the training of the learning phase, the parameters of the algorithms are incrementally learned as data are presented, while in the testing phase such parameters do not change, and hence the algorithms can be compared in terms of performance.

The root mean square error (RMSE) of [3, 37] is used to measure the performance and is expressed as

$$\mathrm{RMSE} = \left(\frac{1}{N} \sum_{k=1}^{N} \widetilde{y}^2(k) \right)^{\frac{1}{2}}, \tag{10.23}$$

where $\widetilde{y}(k)$ is the learning error expressed by Eqs. (10.3), (10.12), and (10.16).

4.1 Experiment 1

Here a real dataset of brain signals consisting of 20000 pairs $(u(k), y(k))$ of 20 s is used to train the training, 2000 pairs ($u(k)$ and $y(k)$) for 2 s used to test the learning. The alpha signal is obtained in this chapter because it has more probabilities to be found. The acquisition system is applied with a 28-year-old healthy man when his

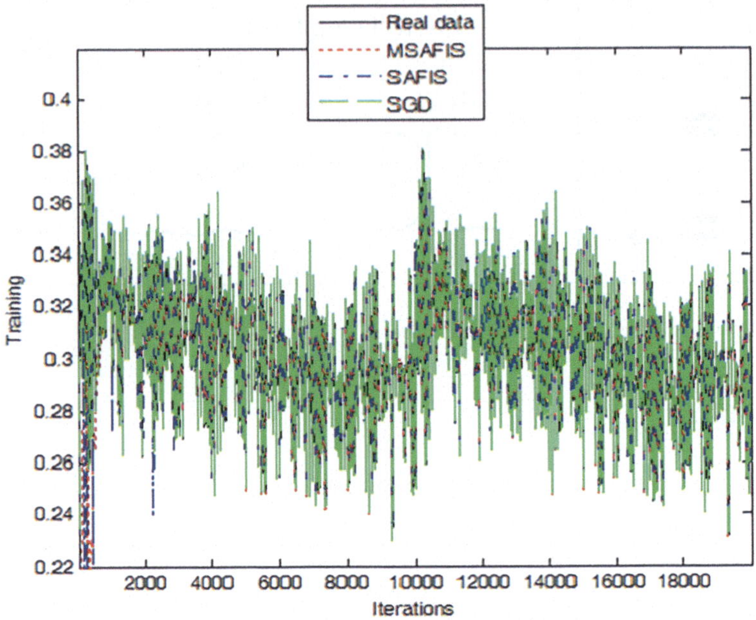

Fig. 10.3 Training for experiment 1

eyes are closed. The inputs of all the intelligent systems are $y(k)$, $y(k+1)$, $y(k+2)$, $y(k+3)$, and the output of the intelligent systems is $y(k+4)$.

Considering Remark 10.3, the parameters for the SAFIS algorithm [35] are $N = 4$, $\tau = 0.99$, $K = 2$, $\epsilon_{\max} = 1$, $\epsilon_{\min} = 0.1$, $y_g = 0.01$, $y_p = 0.001$, $q = 0.1$, $p = 0.1$. Considering Remark 10.7, the parameters of the SGD algorithm of [3] are $N = 4$, $M = 5$, $\alpha_0 = 0.5$. Considering Remark 10.8, the parameters of the MSAFIS are $N = 4$, $\tau = 0.99$, $K = 2$, $\epsilon_{\max} = 2$, $\epsilon_{\min} = 0.2$, $y_g = 0.05$, $y_p = 0.005$, $\alpha_0 = 1$.

Figure 10.3 shows the comparison results for the training of learning in the three algorithms. Figure 10.4 introduces the illustration of the rule evolution for the three algorithms during training. Figure 10.5 presents the comparison results for the testing of learning in the three algorithms. Table 10.3 shows the RMSE comparison results for the algorithms using (10.23).

From Figs. 10.3, 10.4, and 10.5 and Table 10.3, it can be seen that the SGD presents the smallest training RMSE, the MSAFIS presents the smallest testing RMSE, and the MSAFIS obtains the smallest number of rules.

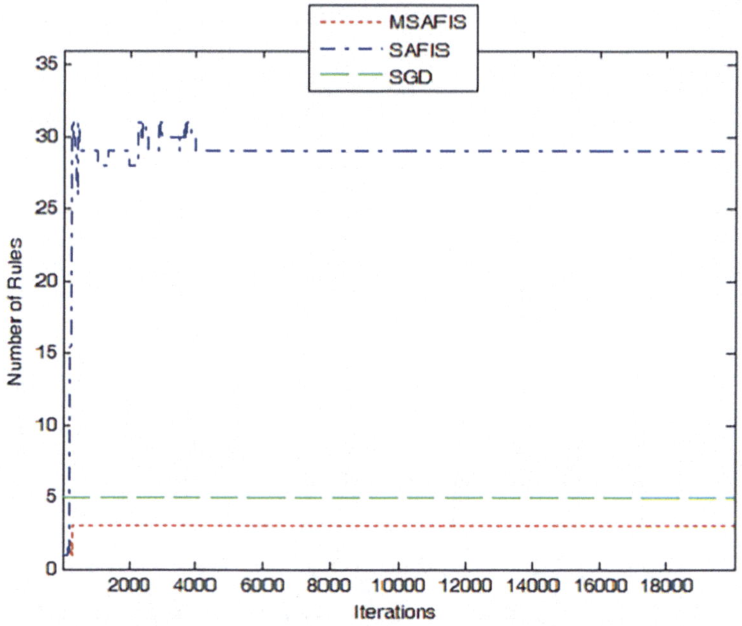

Fig. 10.4 Rule evolution for experiment 1

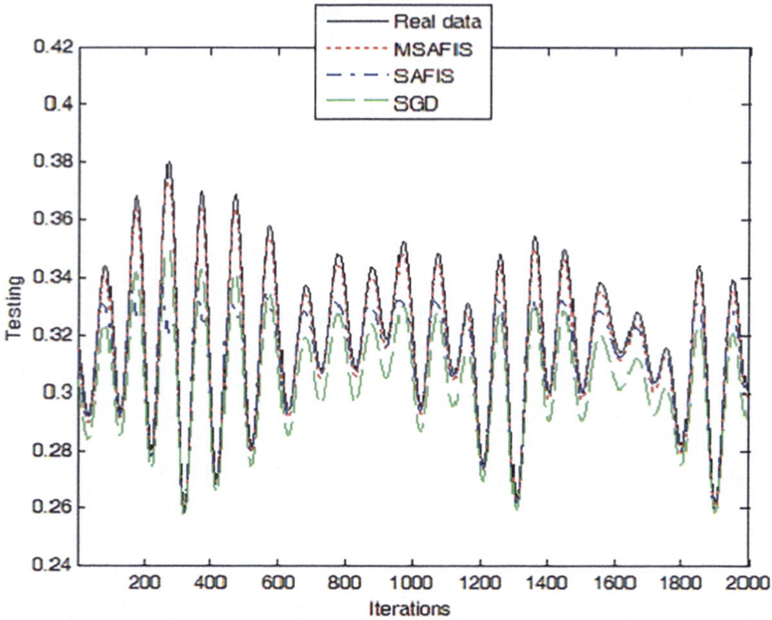

Fig. 10.5 Testing for experiment 1

Table 10.3 Results for experiment 1

Methods	Rules	Training RMSE	Testing RMSE
SGD	5	0.0043	0.0217
SAFIS	29	0.0145	0.0177
MSAFIS	3	0.0331	0.0045

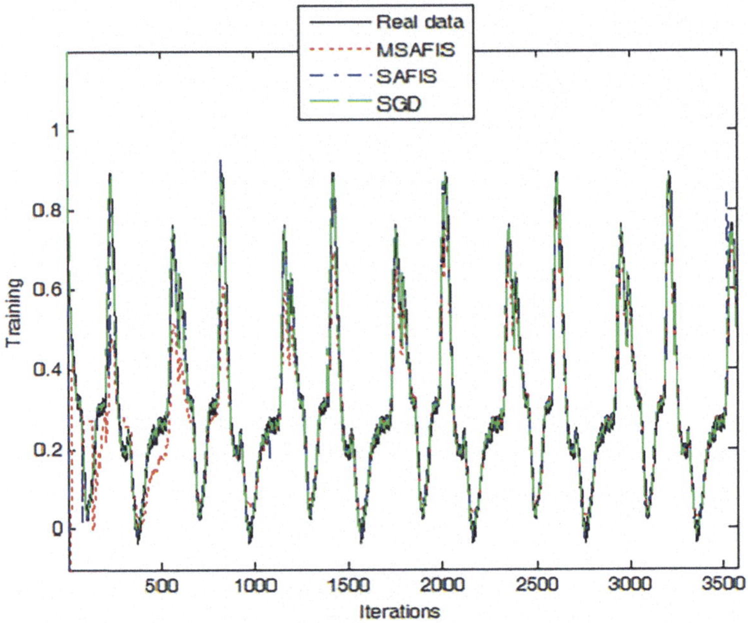

Fig. 10.6 Training for experiment 2

4.2 *Experiment 2*

Here a dataset of eye signals of the down behavior is considered where 3572 pairs $(u(k), y(k))$ of 3.572 s are used to train the learning and 1192 pairs $(u(k), y(k))$ for 1.192 s are used to test the learning. The acquisition system is applied with a 25-year-old healthy man when his eyes are moving, and two electrodes are used to find the signals as described in the aforementioned section. The inputs of all the intelligent systems are $y(k)$, $y(k+1)$, $y(k+2)$, $y(k+3)$, and the output of the intelligent systems is $y(k+4)$.

Considering Remark 10.3, the parameters for the SAFIS [35] are $N = 4$, $\tau = 0.986$, $K = 2$, $\epsilon_{\max} = 2$, $\epsilon_{\min} = 0.2$, $y_g = 0.01$, $y_p = 0.001$, $q = 0.1$, $p = 0.1$. Considering Remark 10.7, the parameters of the SGD [3] are $N = 4$, $M = 9$, $\alpha_0 = 0.5$. Considering Remark 10.8, the parameters of the MSAFIS are $N = 4$, $\tau = 0.986$, $K = 2$, $\epsilon_{\max} = 2$, $\epsilon_{\min} = 0.2$, $y_g = 0.01$, $y_p = 0.001$, $\alpha_0 = 1$.

Figure 10.6 shows the comparison results for the training of learning in the three algorithms. Figure 10.7 introduces the illustration of the rule evolution for the three algorithms during training. Figure 10.8 presents the comparison results for the

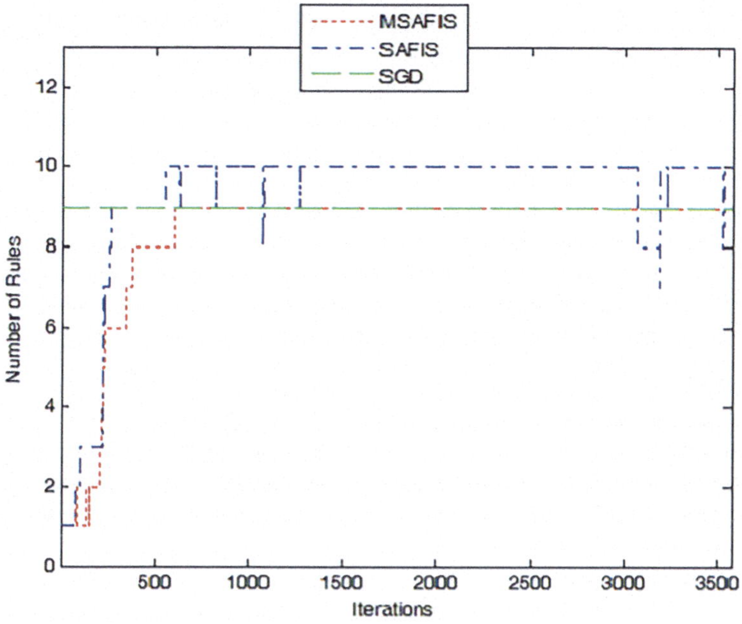

Fig. 10.7 Rule evolution for experiment 2

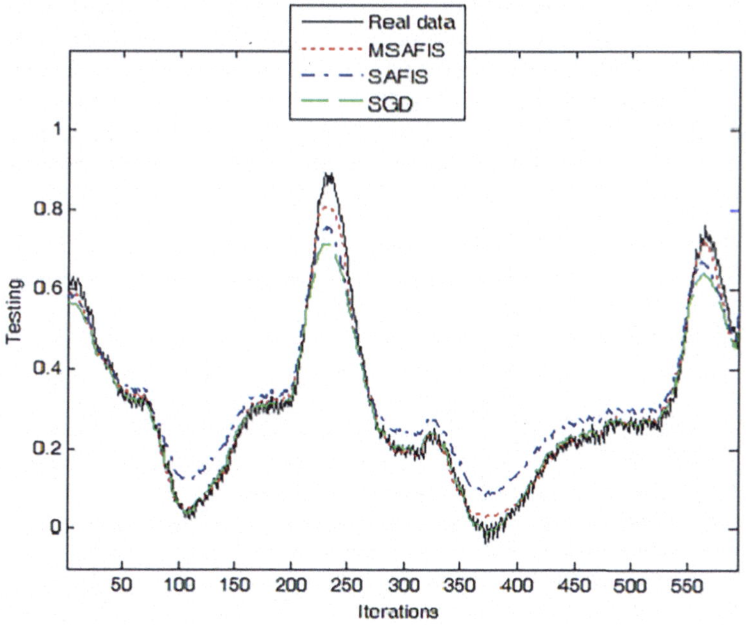

Fig. 10.8 Testing for experiment 2

Table 10.4 Results for experiment 2

Methods	Rules	Training RMSE	Testing RMSE
SGD	9	0.0252	0.0290
SAFIS	10	0.0263	0.0404
MSAFIS	9	0.0706	0.0172

testing of learning in the three algorithms. Table 10.4 shows the RMSE comparison results for the algorithms using (10.23).

From Figs. 10.6, 10.7, and 10.8 and Table 10.4, it can be seen that the SGD presents the smallest training RMSE, the MSAFIS presents the smallest testing RMSE, and the MSAFIS and SGD obtain the smallest number of rules.

Remark 10.9 The SAFIS algorithm is applied in two synthetic examples and in the Makey-Glass time series prediction problem [35]. The SGD algorithm is applied in a synthetic example and in the prediction of the loads distribution in a warehouse [3]. This chapter is novel because it shows that the three algorithms can be used for the learning of other different kind of systems which are the real brain and eye signals with big data.

5 Concluding Remarks

This chapter proposed a combination of two algorithms SAFIS and SGD resulting in MSAFIS. Considering the different experiments, this new algorithm provides better compactness and higher accuracy compared to the original ones. It is worthwhile to mention, because as MSAFIS and SAFIS and SGD are based on online learning, they can handle big datasets of any size. They can also be applied to control, prediction, classification, and diagnosis. Here they were successfully used to learn from a challenging dataset of brain and eye signals. As a future work, the stability of the MSAFIS will be analyzed.

References

1. J. Gama, I. Zliobaite, A. Bifet, M. Pechenizkiy, A. Bouchachia, A survey on concept drift adaptation. ACM Comput. Surv. **46**(4), 1–37 (2014)
2. A. Bouchachia, C. Vanaret, GT2FC: an online growing interval type-2 self-learning fuzzy classifier. IEEE Trans. Fuzzy Syst. **22**(4), 999–1018 (2014)
3. J.J. Rubio, P. Angelov, J. Pacheco, An uniformly stable backpropagation algorithm to train a feedforward neural network. IEEE Trans. Neural Netw. **22**(3), 356–366 (2011)
4. R.-E. Precup, M.C. Sabau, E.M. Petriu, Nature-inspired optimal tuning of input membership functions of Takagi-Sugeno-Kang fuzzy models for anti-lock braking systems. Appl. Soft Comput. **27**, 575–589 (2015)
5. P. Angelov, D. Filev, N. Kasabov, *Evolving Intelligent Systems - Methodology and Applications* (John Wiley & Sons, New York, 2010)

6. A. Bouchachia, *Incremental Learning*. Encyclopedia of Data Warehousing and Mining (2008), pp. 1006–1012
7. J. Gama, *Knowledge Discovery from Data Streams* (Chapman & Hall/CRC, Boca Raton, 2010)
8. N. Kasabov, *Evolving Connectionist Systems: The Knowledge Engineering Approach*, 2nd edn. (Springer Verlag, London, 2007)
9. E. Lughofer, *Evolving Fuzzy Systems - Methodologies*. Advanced Concepts and Applications (Springer, Berlin, 2011)
10. M. Sayed-Mouchaweh, E. Lughofer, *Learning in Non-Stationary Environments: Methods and Applications* (Springer, New York, 2012)
11. A. Bouchachia, Online data processing. Neurocomputing **126**, 116–117 (2014)
12. A. Bouchachia, A. Lena, C. Vanaret, Online and interactive self-adaptive learning of user profile using incremental evolutionary algorithms. Evol. Syst. **5**, 143–157 (2014)
13. J.A. Iglesias, A. Ledezma, A. Sanchis, Evolving classification of Unix users' behaviors. Evol. Syst. **5**, 231–238 (2014)
14. F. Bordignon, F. Gomide, Uninorm based evolving neural networks and approximation capabilities. Neurocomputing **127**, 13–20 (2014)
15. E. Garcia-Cuesta, J.A. Iglesias, User modeling: through statistical analysis and subspace learning. Expert Syst. Appl. **39**, 5243–5250 (2012)
16. F.J. Ordoñez, J.A. Iglesias, P. de Toledo, A. Ledezma, A. Sanchis, Online activity recognition using evolving classifiers. Expert Syst. Appl. **40**, 1248–1255 (2013)
17. F. Gomide, E. Lughofer, Recent advances on evolving intelligent systems and applications. Evol. Syst. **5**, 217–218 (2014)
18. J.A. Iglesias, I. Skrjanc, Applications, results and future direction. Evol. Syst. **5**, 1–2 (2014)
19. E. Lughofer, M. Sayed-Mouchaweh, Adaptive and on-line learning in non-stationary environments. Evol. Syst. **6**, 75–77 (2015)
20. L. Hartert, M. Sayed-Mouchaweh, Dynamic supervised classification method for online monitoring in non-stationary environments. Neurocomputing **126**, 118–131 (2014)
21. J.A. Iglesias, A. Tiemblo, A. Ledezma, A. Sanchis, Web news mining in an evolving framework. Inf. Fusion **28**, 90–98 (2016)
22. G. Klancar, I. Skrjanc, Evolving principal component clustering with a low run-time complexity for IRF data mapping. Appl. Soft Comput. **35**, 349–358 (2015)
23. E. Lughofer, M. Sayed-Mouchaweh, Autonomous data stream clustering implementing split-and-merge concepts - towards a plug-and-play approach. Inf. Sci. **304**, 54–79 (2015)
24. E. Lughofer, C. Cernuda, S. Kindermann, M. Pratama, Generalized smart evolving fuzzy systems. Evol. Syst. **6**, 269–292 (2015)
25. M. Pratama, S.G. Anavatti, M.J. Er, E.D. Lughofer, pClass: an effective classifier for streaming examples. IEEE Trans. Fuzzy Syst. **23**(2), 369–386 (2015)
26. A. Marques Silva, W. Caminhas, A. Lemos, F. Gomide, A fast learning algorithm for evolving neo-fuzzy neuron. Appl. Soft Comput. **14**, 194–209 (2014)
27. M. Sayed-Mouchaweh, E. Lughofer, Decentralized fault diagnosis approach without a global model for fault diagnosis of discrete event systems. Int. J. Control **88**(11), 2228–2241 (2015)
28. C.K. Ahn, A new solution to the induced $l\infty$ finite impulse response filtering problem based on two matrix inequalities. Int. J. Control **87**(2), 404–409 (2014)
29. C.K. Ahn, M.T. Lim, Model predictive stabilizer for T-S fuzzy recurrent multilayer neural network models with general terminal weighting matrix. Neural Comput. Appl. **23**(Suppl. 1), S271–S277 (2013)
30. C.K. Ahn, An error passivation approach to filtering for switched neural networks with noise disturbance. Neural Comput. Appl. **21**(5), 853–861 (2012)
31. J.H. Perez-Cruz, J.J. Rubio, J. Pacheco, E. Soriano, State estimation in MIMO nonlinear systems subject to unknown deadzones using recurrent neural networks. Neural Comput. Appl. **25**(3–4), 693–701 (2014)
32. J.H. Perez-Cruz, J.J. Rubio, R. Encinas, R. Balcazar, Singularity-free neural control for the exponential trajectory tracking in multiple-input uncertain systems with unknown deadzone nonlinearities. Sci. World J. **2014**, 1–10 (2014)

33. C. Torres, J.J. Rubio, C. Aguilar-Ibañez, J.H. Perez-Cruz, Stable optimal control applied to a cylindrical robotic arm. Neural Comput. Appl. **24**(3–4), 937–944 (2014)
34. A. Zdesar, D. Dovzan, I. Skrjanc, Self-tuning of 2 DOF control based on evolving fuzzy model. Appl. Soft Comput. **19**, 403–418 (2014)
35. H.J. Rong, N. Sundararajan, G.B. Huang, P. Saratchandran, Sequential adaptive fuzzy inference system (SAFIS) for nonlinear system identification and prediction. Fuzzy Sets Syst. **157**(9), 1260–1275 (2006)
36. G.B. Huang, P. Saratchandran, N. Sundararajan, An efficient sequential learning algorithm for growing and pruning RBF (GAP-RBF) networks. IEEE Trans. Syst. Man Cybern. B Cybern. **34**(6), 2284–2292 (2004)
37. J.J. Rubio, D.M. Vazquez, D. Mujica-Vargas, Acquisition system and approximation of brain signals. IET Sci. Measure. Technol. **7**(4), 232–239 (2013)
38. J.J. Rubio, F. Ortiz, C.R. Mariaca, J.C. Tovar, A method for online pattern recognition for abnormal eye movements. Neural Comput. Appl. **22**(3–4), 597–605 (2013)

Chapter 11
Error Convergence Analysis of the SAFIS and MSAFIS

1 Introduction

Evolving intelligent networks are inspired by the idea of network model evolution in a dynamically changing and evolving environment. They use gradual change with the aim of life-long modeling and updating self-organization including network structure evolution to update to the environment as structures for information representation with the ability to fully update their structure and adjust their weights. Evolving intelligent networks have been highly applied in prognostic health management plants; two examples are the studying machine failure detection and management of their life cycle.

Evolving intelligent networks have become very popular in the application of prognostic health management plants. Online active modeling concepts have been studied in [1]. In [2], a generalized smart evolving modeling engine of a fuzzy network is investigated. A novel bi-criteria active modeling approach is mentioned in [3]. In [4], a novel incremental type-2 metacognitive extreme modeling machine is addressed. The metacognitive scaffolding modeling machine is introduced in [5]. In [6], a parsimonious random vector functional link network is discussed. A new modeling strategy termed as GenSparseFIS is researched in [7]. In [8] and [9], hybrid dynamic data-driven approaches are suggested. An enhanced convolutional neural network is studied in [10]. A sequential adaptive fuzzy inference system called SAFIS is developed in [11]. In [12], the performance evaluation of the SAFIS is studied. A modified sequential adaptive fuzzy inference system called MSAFIS is proposed in [13]. In evolving intelligent networks, the error convergence is not frequently analyzed.

When the error of a prognostic health management plant is not convergent, the plant output may be infinite even though the plant input is finite. This causes a number of practical problems. For instance, error not convergent in failure detection of a robot arm may cause the robot to move dangerously without any alarm. Also, these errors that are not convergent often incur a certain amount of physical

J. de J. Rubio, *Stability Analysis of Neural Networks and Evolving Intelligent Systems*, https://doi.org/10.1007/978-3-031-87282-2_11

damage in plants, which can become costly. Nonetheless, errors are inherently not convergent in many plants, for example, a fighter jet or a rocket at liftoff. Although evolving intelligent networks can be designed to be applied in prognostic health management plants, it is important to ensure their error convergence to reach an acceptable performance. Error convergent intelligent networks are characterized by the boundedness criterion, i.e., if bounded inputs are utilized, then outputs are ensured to be bounded.

Error convergent intelligent networks also have become very popular in the application of prognostic health management plants. The L-infinity performance analysis of a neural network is taken into account in [14]. In [15] and [16], robust evolving cloud-based controllers are presented. Robust common spatial pattern feature extraction algorithms are designed in [17] and [18]. In [19], a command-filtered backstepping update control is researched. Composite update locally weighted modeling control approaches are proposed in [20] and [21]. In [22] and [23], the asymptotic error convergence analysis of generalized neural networks is addressed. Fuzzy convergent controllers are discussed in [24] and [25]. It is not frequent that error convergent intelligent networks are also evolving.

In this chapter, Lyapunov strategy is utilized to analyze error convergence of the SAFIS and MSAFIS for their application in prognostic health management plants. SAFIS employs an extended Kalman filter, and it is linearized to get its modeling dynamic equation; after, error convergence based on the Lyapunov strategy is analyzed. MSAFIS employs the gradient descent technique, and it is linearized to acquire its modeling dynamic equation; later, error convergence based on the Lyapunov strategy is analyzed. Figure 11.1 shows the error convergence analysis steps of both algorithms.

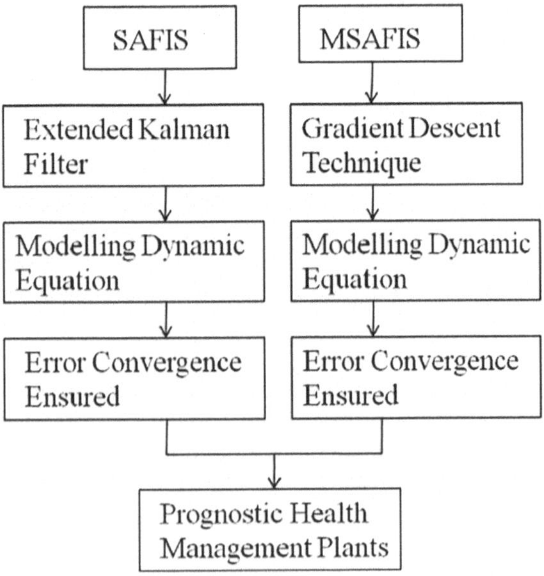

Fig. 11.1 Error convergence analysis of the SUFIN and CSUFIN

The chapter is organized as follows. In Sect. 3, error convergence of the SAFIS is analyzed. Error convergence of the MSAFIS is analyzed in Sect. 4. In Sect. 5, performance of the SAFIS and MSAFIS is detailed in two examples. The conclusion and future research are explained in Sect. 6.

2 Prognostic Health Management Plant

Take into account the next prognostic health management plant:

$$\gamma(k) = f\left[\chi(k)\right], \tag{11.1}$$

where $\chi(k) = [\chi_1(k), \ldots, \chi_i(k), \ldots, \chi_N(k)]^T = [\gamma(k-1), \ldots, \gamma(k-n), \upsilon(k-1), \ldots, \upsilon(k-m)]^T \in \Re^{N\times 1}$ $(N = n + m)$ is the input vector, $\upsilon(k-1) \in \Re$ is the plant input, $\gamma(k) \in \Re$ is the plant output, and f is an unknown nonlinear function, $f \in C^{\infty}$.

3 Error Convergence Analysis of the SAFIS

3.1 Description of the SAFIS

The sequential adaptive fuzzy inference system (SAFIS) is developed based on the functional equivalence between a radial basis function network and a fuzzy inference network producing a fuzzy neural network. In SAFIS, using that neurons are added or removed based on the input data received so far. If the input data do not warrant adding of neurons, then only weights of the "closest" (in a Euclidean sense) neuron are updated using an extended Kalman filter.

The SAFIS algorithm is summarized in the next paragraphs [11].

For each data $(\chi(k), \gamma(k))$, where $\chi(k) \in \Re^N$, $\gamma(k) \in \Re$, and $k = 1, 2, \ldots,$ do:

(1) Get the network output:

$$\begin{aligned} \widehat{\gamma}_s(k) &= \frac{\sum_{r=1}^{L_s} o_{s,r}(k) S_{s,r}(k)}{\sum_{r=1}^{L_s} S_{s,r}(k)}, \\ S_{s,r}(k) &= \exp\left(-\frac{1}{\xi_{s,r}^2(k)} \left\| \chi_i(k) - \mu_{s,r}(k) \right\|^2\right), \end{aligned} \tag{11.2}$$

where L_s is the neuron number, $S_{s,r}(k)$ is the firing strength of the rth neuron, and $o_{s,r}(k)$ is the normalized neuron. Note that each neuron is expressed as a radial basis function network described by its center $\mu_{s,r}(k)$ and its spread $\xi_{s,r}(k)$.

(2) Acquire terms needed in the growth criterion:

$$\varepsilon_s(k) = \max\left\{\varepsilon_{s,\max}\eta_s^k, \varepsilon_{s,\min}\right\}, 0 < \eta_s < 1, \tag{11.3}$$

where $\varepsilon_{s,\max}$ and $\varepsilon_{s,\min}$ are the threshold largest and smallest distances admitted between inputs and the corresponding nearest center of neurons. The term η_s ($0 < \eta_s < 1$) indicates the decay constant. The modeling error is in the next equation:

$$\widetilde{\gamma}_s(k) = \widehat{\gamma}_s(k) - \gamma(k), \tag{11.4}$$

where $\gamma(k)$ and $\widehat{\gamma}_s(k)$ are the output and estimated output, respectively.

(3) Use the criterion for adding neurons if the next two conditions are fulfilled:

If

$$\left\|\chi_i(k) - \mu_{s,r}(k)\right\| > \varepsilon_s(k), \tag{11.5}$$

and

$$Y_{s,\inf}(L_s + 1) = |\widetilde{\gamma}_s(k)| \frac{\left(1.8K_s \left\|\chi_i(k) - \mu_{s,r}(k)\right\|\right)^N}{\sum_{r=1}^{L_s+1}(1.8\xi_{s,r}(k))^N} > \gamma_{s,g}, \tag{11.6}$$

where $\gamma_{s,g}$ is the growing threshold. A new neuron $L_s + 1$ is added if $\gamma_{s,g}$ is exceeded.

The new neuron $L_s + 1$ is in the next equation:

$$\begin{aligned} o_{s,L_s+1}(k) &= -\widetilde{\gamma}_s(k), \\ \mu_{s,L_s+1}(k) &= \chi_i(k), \\ \xi_{s,L_s+1}(k) &= K_s \left\|\chi_i(k) - \mu_{s,L_s+1}(k)\right\|. \end{aligned} \tag{11.7}$$

If no neuron is added, the nearest neuron rs is gotten in the next equation:

$$\min_r S_{s,r}(k) \Longrightarrow rs = r, \tag{11.8}$$

and update the network weights $o_{s,r}(k)$, $\mu_{s,r}(k)$, $\xi_{s,r}(k)$ for the nearest neuron by using the extended Kalman filter:

$$\begin{aligned}\varphi_s(k+1) &= \varphi_s(k) - \frac{1}{a_s(k)}G_s(k+1)d_s(k)\widetilde{\gamma}_s(k),\\ G_s(k+1) &= G_s(k) - \frac{1}{b_s(k)}G_s(k)d_s(k)d_s^T(k)G_s(k),\end{aligned} \tag{11.9}$$

with

$$\begin{aligned}\varphi_s(k) &= \left[\varphi_{s,1}(k), \varphi_{s,2}(k), \varphi_{s,3}(k)\right]^T,\\ d_s(k) &= \left[d_{s,1}(k), d_{s,2}(k), d_{s,3}(k)\right]^T,\end{aligned}$$

and

$$\begin{aligned}\varphi_{s,1}(k) &= \mu_{s,rs}(k), \varphi_{s,2}(k) = \xi_{s,rs}(k), \varphi_{s,3}(k) = o_{s,rs}(k),\\ d_{s,1}(k) &= \frac{2[o_{s,rs}(k)-\widehat{\gamma}_s(k)]S_{s,rs}(k)[\chi_i(k)-\mu_{s,rs}(k)]}{\left[\sum_{r=1}^{L_s} S_{s,r}(k)\right]\xi_{s,rs}^2(k)},\\ d_{s,2}(k) &= \frac{2[o_{s,rs}(k)-\widehat{\gamma}_s(k)]S_{s,rs}(k)\|\chi_i(k)-\mu_{s,rs}(k)\|^2}{\left[\sum_{r=1}^{L_s} S_{s,r}(k)\right]\xi_{s,rs}^3(k)},\\ d_{s,3}(k) &= \frac{S_{s,rs}(k)}{\left[\sum_{r=1}^{L_s} S_{s,r}(k)\right]},\end{aligned}$$

$a_s(k) = b_{s,2} + d_s^T(k)G_s(k)d_s(k) \in \Re$, $b_s(k) = a_s(k) + d_s^T(k)G_s(k)d_s(k) \in \Re$, $G_s(1)=g_{s,e}I$, $g_{s,e}$ is a term selected by the designer, $0 < g_{s,e} \le 1$, $0 < b_{s,2} \le 1$, and I is the identity matrix.

(4) If the next criterion is fulfilled:

$$Y_{s,\inf}(rs) = \left|o_{s,rs}(k)\right| \frac{\left(1.8\xi_{s,rs}(k)\right)^N}{\sum_{r=1}^{L_s}(1.8\xi_{s,r}(k))^N} < \gamma_{s,p}, \tag{11.10}$$

then, remove the rs neuron, and reduce the dimensionality of extended Kalman filter. Note that $\gamma_{s,p}$ is the pruning threshold.

3.2 Linearization of the SAFIS

The linearization of SAFIS is needed for its error convergence analysis.

Utilize the SAFIS output of (11.2) in the next equation:

$$\widehat{\gamma}_s(k) = \mathcal{F}_s(k) = \frac{\sum_{r=1}^{L_s} o_{s,r}(k) S_{s,r}(k)}{\sum_{r=1}^{L_s} S_{s,r}(k)}, \quad S_{s,r}(k) = \exp\left(-\frac{1}{\xi_{s,r}^2(k)} \left\| \chi_i(k) - \mu_{s,r}(k) \right\|^2\right). \tag{11.11}$$

According to the Stone-Weierstrass theorem, the unknown nonlinear function f of (11.1) is approximated in the next equation:

$$\gamma(k) = \mathcal{F}_{s,*} + \in_{s,f} = \frac{\sum_{r=1}^{L_s} o_{s,r*} S_{s,r*}(k)}{\sum_{r=1}^{L_s} S_{s,r*}(k)} + \in_{s,f}, \quad S_{s,r*}(k) = \exp\left(-\frac{1}{\xi_{s,r*}^2} \left\| \chi_i(k) - \mu_{s,r*} \right\|^2\right), \tag{11.12}$$

where $\in_{s,f} = \gamma(k) - \mathcal{F}_{s,*} \in \Re$ is the modeling error, $S_{s,r*}(k) \in \Re$, $\mu_{s,r*} \in \Re$, $\xi_{s,r*} \in \Re$, $o_{s,r*} \in \Re$, $\mu_{s,r*}$, $\xi_{s,r*}$, and $o_{s,r*}$ are the optimal weights that can minimize the modeling error $\in_{s,f}$. In the case of three independent variables, a function has a Taylor series of the next equation:

$$\begin{aligned} f(\varpi_1, \varpi_2, \varpi_3) &= f(\varpi_{10}, \varpi_{20}, \varpi_{30}) \\ &+ (\varpi_1 - \varpi_{10}) \frac{\partial f(\varpi_1, \varpi_2, \varpi_3)}{\partial \varpi_1} + (\varpi_2 - \varpi_{20}) \frac{\partial f(\varpi_1, \varpi_2, \varpi_3)}{\partial \varpi_2} \\ &+ (\varpi_3 - \varpi_{30}) \frac{\partial f(\varpi_1, \varpi_2, \varpi_3)}{\partial \varpi_3} + b_{s,f}, \end{aligned} \tag{11.13}$$

where $b_{s,f} \in \Re$ is the remainder of the Taylor series. ϖ_1, ϖ_2, and ϖ_3 correspond to $\mu_{s,r}(k) \in \Re$, $\xi_{s,r}(k) \in \Re$, and $o_{s,r}(k) \in \Re$, ϖ_{10}, ϖ_{20}, and ϖ_{30} correspond to $\mu_{s,r*} \in \Re$, $\xi_{s,r*} \in \Re$, and $o_{s,r*} \in \Re$; therefore, the Taylor series is applied to linearize (11.11) as in the next equation:

$$\begin{aligned} \mathcal{F}_s(k) &= \mathcal{F}_{s,*} + \sum_{r=1}^{L_s} \widetilde{\mu}_{s,r}(k) \frac{\partial \mathcal{F}_s(k)}{\partial \mu_{s,r}(k)} \\ &+ \sum_{r=1}^{L_s} \widetilde{\xi}_{s,r}(k) \frac{\partial \mathcal{F}_s(k)}{\partial \xi_{s,r}(k)} + \sum_{r=1}^{L_s} \widetilde{o}_{s,r}(k) \frac{\partial \mathcal{F}_s(k)}{\partial o_{s,r}(k)} + b_{s,f}, \end{aligned} \tag{11.14}$$

where $\widetilde{\mu}_{s,r}(k) = \mu_{s,r}(k) - \mu_{s,r*} \in \Re$, $\widetilde{\xi}_{s,r}(k) = \xi_{s,r}(k) - \xi_{s,r*} \in \Re$, and $\widetilde{o}_{s,r}(k) = o_{s,r}(k) - o_{s,r*} \in \Re$. Acquiring partial derivatives, it produces

$$\begin{aligned}\frac{\partial \mathcal{F}_s(k)}{\partial \mu_{s,r}(k)} &= d_{s,1}(k)\\ &= \frac{2\left[o_{s,rs}(k)-\widehat{\gamma}_s(k)\right]S_{s,rs}(k)\left[\chi_i(k)-\mu_{s,rs}(k)\right]}{\left[\sum_{r=1}^{L_s} S_{s,r}(k)\right]\xi^2_{s,rs}(k)}.\end{aligned} \tag{11.15}$$

Subsequently,

$$\begin{aligned}\frac{\partial \mathcal{F}_s(k)}{\partial \xi_{s,r}(k)} &= d_{s,2}(k)\\ &= \frac{2\left[o_{s,rs}(k)-\widehat{\gamma}_s(k)\right]S_{s,rs}(k)\left\|\chi_i(k)-\mu_{s,rs}(k)\right\|^2}{\left[\sum_{r=1}^{L_s} S_{s,r}(k)\right]\xi^3_{s,rs}(k)},\end{aligned} \tag{11.16}$$

and

$$\frac{\partial \mathcal{F}_s(k)}{\partial o_{s,r}(k)} = d_{s,3}(k) = \frac{S_{s,rs}(k)}{\left[\sum_{r=1}^{L_s} S_{s,r}(k)\right]}. \tag{11.17}$$

Substituting $d_{s,1}(k)$ of (11.15), $d_{s,2}(k)$ of (11.16), and $d_{s,3}(k)$ of (11.17) into (11.14), it produces

$$\begin{aligned}\mathcal{F}_s(k) &= \mathcal{F}_{s,*} + \sum_{r=1}^{L_s} \widetilde{\mu}_{s,r}(k) d_{s,1}(k)\\ &+ \sum_{r=1}^{L_s} \widetilde{\xi}_{s,r}(k) d_{s,2}(k) + \sum_{r=1}^{L_s} \widetilde{o}_{s,r}(k) d_{s,3}(k) + b_{s,f}.\end{aligned} \tag{11.18}$$

Take into account the modeling error $\widetilde{\gamma}_s(k) \in \Re$ of (11.4) of the next equation:

$$\widetilde{\gamma}_s(k) = \widehat{\gamma}_s(k) - \gamma(k), \tag{11.19}$$

where $\gamma(k)$ and $\widehat{\gamma}_s(k)$ are defined in (11.1) and (11.11), respectively. Substituting (11.11), (11.12), and (11.19) into (11.18) produces

$$\begin{aligned}\widetilde{\gamma}_s(k) &= \sum_{r=1}^{L_s}\widetilde{\mu}_{s,r}(k)d_{s,1}(k) + \sum_{r=1}^{L_s}\widetilde{\xi}_{s,r}(k)d_{s,2}(k)\\ &+\sum_{r=1}^{L_s}\widetilde{o}_{s,r}(k)d_{s,3}(k) + \beta_s(k),\end{aligned} \tag{11.20}$$

where $\beta_s(k) = b_{s,f} - \epsilon_{s,f}$.

From (11.20), the modeling dynamic equation can be expressed as in the next equation:

$$\widetilde{\gamma}_s(k) = d_s^T(k)\widetilde{\varphi}_s(k) + \beta_s(k), \tag{11.21}$$

where $d_s(k) = [d_{s,1}(k), d_{s,2}(k), d_{s,3}(k)]^T \in \Re^{1\times 3L_s}$, $\widetilde{\varphi}_s(k) = \left[\widetilde{\varphi}_{s,1}(k), \widetilde{\varphi}_{s,2}(k), \widetilde{\varphi}_{s,3}(k)\right]^T = [\widetilde{\mu}_{s,rs}(k), \widetilde{\xi}_{s,rs}(k), \widetilde{o}_{s,rs}(k)]^T \in \Re^{3L_s\times 1}$. From $\widetilde{\mu}_{s,r}(k)$, $\widetilde{\xi}_{s,r}(k)$, and $\widetilde{o}_{s,r}(k)$ of (11.14) produces $\widetilde{\varphi}_s(k) = \varphi_s(k) - \varphi_{s,*}$, $\varphi_{s,*}$ are the optimal weights that can minimize the modeling error $\beta_s(k)$.

3.3 Error Convergence of the SAFIS

In this section, the error convergence of the SAFIS is analyzed. Lyapunov strategy is selected because it can be used for the error convergence analysis of nonlinear networks. The next theorem shows the first main contribution of this chapter.

Theorem 11.1 *The modeling error of the extended Kalman filter (11.4), (11.9) as updating of the SAFIS (11.2), (11.11) applied for the modeling of prognostic health management plants (11.1) is uniformly convergent, and the upper bound of the average modeling error $\Omega_s(k)$ fulfills*

$$\limsup_{T\to\infty}\frac{1}{T}\sum_{k=2}^{T}\Omega_s(k) \le \frac{\overline{\beta}_s^2}{b_{s,2}}, \tag{11.22}$$

where $\Omega_s(k) = \frac{\left[d_s^T(k)G_s(k)d_s(k)\right]^2}{b_s(k)a_s^2(k)}\widetilde{\gamma}_s^2(k)$, $a_s(k) = b_{s,2} + d_s^T(k)G_s(k)d_s(k) > 0$, $b_s(k) = a_s(k) + d_s^T(k)G_s(k)d_s(k) > 0$.

Proof See [26] for the proof.

4 Error Convergence Analysis of the MSAFIS

4.1 *Description of the MSAFIS*

The modified sequential adaptive fuzzy inference system (MSAFIS) is the SAFIS algorithm with the modification of the extended Kalman filter (11.9) by the gradient descent technique, but using the similar structure than the SAFIS algorithm.

The MSAFIS algorithm is summarized in [13]. The MSAFIS algorithm basically utilizes Eqs. (11.2)–(11.8), (11.10) of the SAFIS algorithm. The difference of the MSAFIS is in Eq. (11.9), where the SAFIS uses the extended Kalman filter and it is denoted with a subscript s, while the MSAFIS uses the gradient descent technique and it is denoted with a subscript c. The change is detailed in the next sentence.

Update network weights $o_{c,r}(k)$, $\mu_{c,r}(k)$, $\xi_{c,r}(k)$ for the nearest neuron by using the gradient descent technique:

$$\begin{aligned} \varphi_c(k+1) &= \varphi_c(k) - g_c(k) d_c(k) \widetilde{\gamma}_c(k), \\ g_c(k) &= \frac{g_{c,g}}{2\left(\frac{1}{2} + \sum_{j=1}^{3} d_{c,j}^2(k)\right)}, \end{aligned} \tag{11.23}$$

with

$$\begin{aligned} \varphi_c(k) &= \left[\varphi_{c,1}(k), \varphi_{c,2}(k), \varphi_{c,3}(k)\right]^T, \\ d_c(k) &= \left[d_{c,1}(k), d_{c,2}(k), d_{c,3}(k)\right]^T, \end{aligned}$$

and

$$\begin{aligned} \varphi_{c,1}(k) &= \mu_{c,rc}(k), \ \varphi_{c,2}(k) = \xi_{c,rc}(k), \ \varphi_{c,3}(k) = o_{c,rc}(k), \\ d_{c,1}(k) &= \frac{2\left[o_{c,rc}(k) - \widehat{\gamma}_c(k)\right] S_{c,rc}(k) \left[\chi_i(k) - \mu_{c,rc}(k)\right]}{\left[\sum_{r=1}^{L_c} S_{c,r}(k)\right] \xi_{c,rc}^2(k)}, \\ d_{c,2}(k) &= \frac{2\left[o_{c,rc}(k) - \widehat{\gamma}_c(k)\right] S_{c,rc}(k) \left\| \chi_i(k) - \mu_{c,rc}(k) \right\|^2}{\left[\sum_{r=1}^{L_c} S_{c,r}(k)\right] \xi_{c,rc}^3(k)}, \\ d_{c,3}(k) &= \frac{S_{c,rc}(k)}{\left[\sum_{r=1}^{L_c} S_{c,r}(k)\right]}, \end{aligned}$$

$g_c(k)$ is the time varying rate, and $g_{c,g}$ is a term selected by the designer, $0 < g_{c,g} \leq 1$.

Remark 11.1 In SAFIS and MSAFIS, some terms should be selected in advance according to the prognostic health management plant. They include the distance thresholds ($\varepsilon_{s,\max}$, $\varepsilon_{s,\min}$, η_s) and ($\varepsilon_{c,\max}$, $\varepsilon_{c,\min}$, η_c), the overlap factors K_s and K_c for determining the width of newly added neurons, the growing thresholds ($\gamma_{s,g}$) and ($\gamma_{c,p}$) for adding a new significant neuron, and the pruning thresholds ($\gamma_{s,p}$) and ($\gamma_{c,p}$) for removing an insignificant neuron. A general selection procedure for predefined terms is in the next sentence: max is set to around the upper bound of input variables, $\varepsilon_{s,\min}$ and $\varepsilon_{c,\min}$ are set to around 10% of $\varepsilon_{s,\max}$ and $\varepsilon_{c,\max}$, and η_s and η_c are set to around 0.99. $\gamma_{s,p}$ and $\gamma_{c,p}$ are set to around 10% of $\gamma_{s,g}$ and $\gamma_{c,g}$. $\varepsilon_{s,\max}$ and $\varepsilon_{c,\max}$ are seen in the range [1.0, 10.0]. The overlap factors K_s and K_c are utilized to initialize the width of the newly added neuron and selected according to the prognostic health management plant, and they are seen in the range [1.0, 2.0]. The growing thresholds $\gamma_{s,g}$ and $\gamma_{c,g}$ are selected according to the network performance, and they are seen in the range [0.001, 0.05]. The smaller the $\gamma_{s,g}$ and $\gamma_{c,g}$, the better the network performance, but the resulting network structures are more complex.

Remark 11.2 In the SAFIS and MSAFIS, each neuron r is equivalent to each rule r, and the neurons number L_s'and L_c are equivalent to the rules number L_s and L_c. Thus, the neurons are equivalent to the rules in the SAFIS and MSAFIS.

Remark 11.3 Even the SAFIS and MSAFIS are in structure similar, they are completely different in the weights adjust; the SAFIS uses the extended Kalman filter (11.9), while the MSAFIS uses the gradient descent technique (11.23), and it produces significant changes in the structures of both algorithms.

Remark 11.4 The SAFIS and MSAFIS have the purpose to use the least required neurons to get a satisfactory modeling; they have one hidden layer with the least possible neurons number. The SAFIS and MSAFIS have two types of scalability: the first scalability is to increase the neurons number in the hidden layer, and the second scalability is to include other hidden layer with more neurons. Both types of scalability are contrary to the main purpose of both algorithms.

Remark 11.5 Since SAFIS and MSAFIS are self-organization algorithms to dynamically update their structure and adjust their weights to get an acceptable modeling, they gradually optimize their weights and structure.

Remark 11.6 Step 4 of the SAFIS and MSAFIS uses pruning algorithms to remove the insignificant neurons; it avoids that SAFIS and MSAFIS grow indeterminately.

4.2 Linearization of the MSAFIS

The linearization of MSAFIS is needed for its error convergence analysis.

Use the MSAFIS output of (11.2) by changing the subscript s for c as in the next equation:

$$\widehat{\gamma}_c(k) = \mathcal{F}_c(k) = \frac{\sum_{r=1}^{L_c} o_{c,r}(k) S_{c,r}(k)}{\sum_{r=1}^{L_c} S_{c,r}(k)}, \qquad (11.24)$$
$$S_{c,r}(k) = \exp\left(-\frac{1}{\xi_{c,r}^2(k)} \left\| \chi_i(k) - \mu_{c,r}(k) \right\|^2\right).$$

Using the same linearization method described by Eqs. (11.11)–(11.20) produces

$$\widetilde{\gamma}_c(k) = \sum_{r=1}^{L_c} \widetilde{\mu}_{c,r}(k) d_{c,1}(k) + \sum_{r=1}^{L_c} \widetilde{\xi}_{c,r}(k) d_{c,2}(k) \qquad (11.25)$$
$$+ \sum_{r=1}^{L_c} \widetilde{o}_{c,r}(k) d_{c,3}(k) + \beta_c(k),$$

where $\beta_c(k) = b_{c,f} - \epsilon_{c,f}$. $\widetilde{\mu}_{c,r}(k) = \mu_{c,r}(k) - \mu_{c,r*} \in \Re$, $\widetilde{\xi}_{c,r}(k) = \xi_{c,r}(k) - \xi_{c,r*} \in \Re$, $\widetilde{o}_{c,r}(k) = o_{c,r}(k) - o_{c,r*} \in \Re$. $\mu_{c,r}(k)$, $\xi_{c,r}(k)$, $o_{c,r}(k)$, $d_{c,1}(k)$, $d_{c,2}(k)$, $d_{c,3}(k)$ are described (11.23). $\mu_{c,r*}$, $\xi_{c,r*}$, $o_{c,r*}$ are the optimal weights that can minimize the modeling error $\beta_c(k)$.

From (11.25), the modeling dynamic equation can be expressed as in the next equation:

$$\widetilde{\gamma}_c(k) = d_c^T(k) \widetilde{\varphi}_c(k) + \beta_c(k), \qquad (11.26)$$

where $d_c(k) = [d_{c,1}(k), d_{c,2}(k), d_{c,3}(k)]^T \in \Re^{1 \times 3L_c}$, $\widetilde{\varphi}_c(k) = \left[\widetilde{\varphi}_{c,1}(k), \widetilde{\varphi}_{c,2}(k), \widetilde{\varphi}_{c,3}(k)\right]^T = [\widetilde{\mu}_{c,rc}(k), \widetilde{\xi}_{c,rc}(k), \widetilde{o}_{c,rc}(k)]^T \in \Re^{3L_c \times 1}$. From $\widetilde{\mu}_{c,r}(k)$, $\widetilde{\xi}_{c,r}(k)$, and $\widetilde{o}_{c,r}(k)$ of (11.25) produces $\widetilde{\varphi}_c(k) = \varphi_c(k) - \varphi_{c,*}$, $\varphi_{c,*}$ are the optimal weights that can minimize the modeling error $\beta_c(k)$.

4.3 Error Convergence of the MSAFIS

In this section, the error convergence of the MSAFIS is analyzed. Lyapunov strategy is selected because it can be used for the error convergence analysis of nonlinear networks. The next theorem shows the second main contribution of this chapter.

Theorem 11.2 *The modeling error of the gradient descent technique (11.4), (11.23) as updating of the MSAFIS (11.24) applied for the modeling of prognostic health management plants (11.1) is uniformly convergent, and the upper bound of the average modeling error* $\Omega_c(k)$ *fulfills*

$$\limsup_{T\to\infty}\frac{1}{T}\sum_{k=2}^{T}\Omega_c(k) \le g_{c,g}\overline{\beta}_c^2, \tag{11.27}$$

where $\Omega_c(k) = \frac{g_c(k-1)}{2}\widetilde{\gamma}_c^2(k-1)$, $0 < g_{c,g} \le 1 \in \Re$ *and* $0 < g_c(k) \in \Re$ *are described in (11.23),* $\widetilde{\gamma}_c(k)$ *are described in (11.4),* $\overline{\beta}_c$ *is the upper bound of the uncertainty* $\beta_c(k)$, $|\beta_c(k)| < \overline{\beta}_c$.

Proof See [26] for the proof.

Remark 11.7 The terms L_s (neurons number) in the SAFIS and L_c (neurons number) in the MSAFIS are finite, because the algorithms add the significant neurons and prune the insignificant neurons to update themself to the changing environment. The neuron numbers L_s and L_c are changed by the adding and pruning algorithms, and L_s and L_c change only the dimension of $d_s^T(k)$, $\varphi_s(k)$, $d_c^T(k)$, and $\varphi_c(k)$; thus, the error convergence results are preserved.

5 Examples

In this part of the chapter, the studied algorithms are applied for the modeling of two numerical examples. The two selected numerical examples have the two main characteristics: First, they are nonlinear plants with the structure of Eq. (11.1), and second, they let to show the characteristics of both algorithms. In all cases, the MSAFIS will be compared with the SAFIS. The differences between three algorithms were explained in before sections. The root mean square error denoted as MSE is utilized for comparisons:

$$\text{MSE} = \left(\frac{1}{T}\sum_{k=1}^{T}\widetilde{\gamma}^2(k)\right)^{\frac{1}{2}}, \tag{11.28}$$

with $\widetilde{\gamma}^2(k) = \widetilde{\gamma}_s^2(k)$ as the modeling error for the SAFIS, and $\widetilde{\gamma}^2(k) = \widetilde{\gamma}_c^2(k)$ as the modeling error for the MSAFIS.

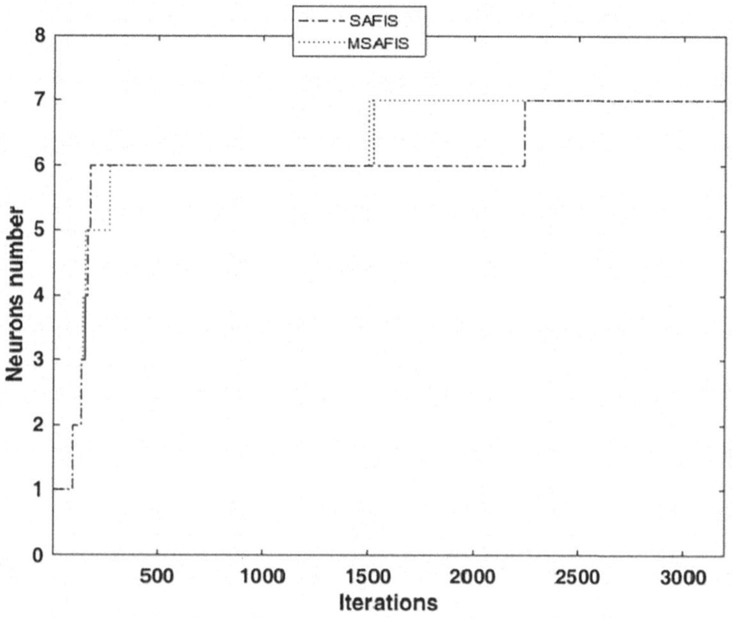

Fig. 11.2 Neurons number for Example 1

5.1 Example 1

The plant of Example 1 is expressed in the next equation:

$$\begin{aligned} \gamma(k) &= \frac{\gamma(k-1)\gamma(k-2)[\gamma(k-1)-0.5]}{1+\gamma(k-1)^2+\gamma(k-2)^2} + \upsilon(k-1), \\ \upsilon(k-1) &= \sin\left(\frac{2\pi(k-1)}{25}\right). \end{aligned} \tag{11.29}$$

The nonlinear plant of Eqs. (11.1) and (11.29) is utilized where inputs are $\chi_1(k) = \gamma(k-1)$, $\chi_2(k) = \gamma(k-2)$, $\chi_3(k) = \upsilon(k-1)$ and the output is $\gamma(k) = \gamma(k)$. The data of 3000 iterations are used for the training.

Terms of the SAFIS algorithm [11] are $N = 3$, $\eta_s = 0.99$, $K_s = 1$, $\varepsilon_{s,\max} = 5$, $\varepsilon_{s,\min} = 0.5$, $\gamma_{s,g} = 0.01$, $\gamma_{s,p} = 0.001$, $g_{s,e} = 0.01$, $b_{s,2} = 0.2$.

Terms of the MSAFIS algorithm [13] are $N = 3$, $\eta_c = 0.99$, $K_c = 1$, $\varepsilon_{c,\max} = 5$, $\varepsilon_{c,\min} = 0.5$, $\gamma_{c,g} = 0.01$, $\gamma_{c,p} = 0.001$, $g_{c,g} = 1$.

Figures 11.2, 11.3, 11.4, and 11.5 show the comparisons for the neurons number, the generated neurons, MSE convergence, and training of the SAFIS and MSAFIS. The training MSE comparisons of (11.28) are shown in Table 11.1.

From Figs. 11.2 and 11.3, it is observed that both algorithms reach the same neurons number. From Fig. 11.4 and Table 11.1, it is observed that the MSAFIS has better convergence than the SAFIS because the MSE is smaller for the first. From Fig. 11.5, it is observed that the MSAFIS improves the SAFIS because the signal

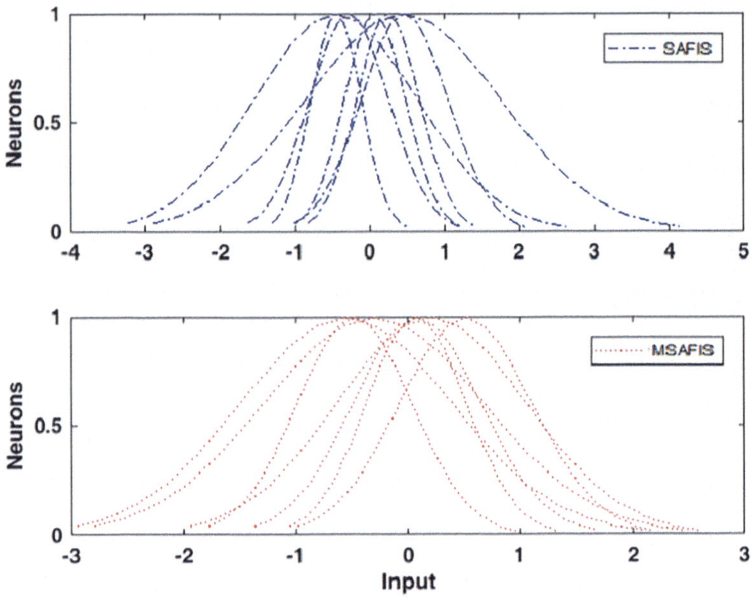

Fig. 11.3 Generated neurons for Example 1

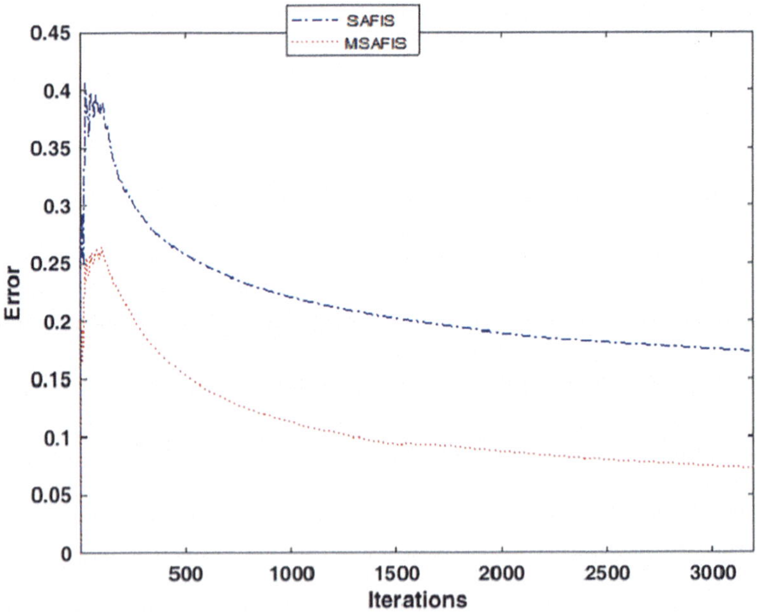

Fig. 11.4 MSE convergence for Example 1

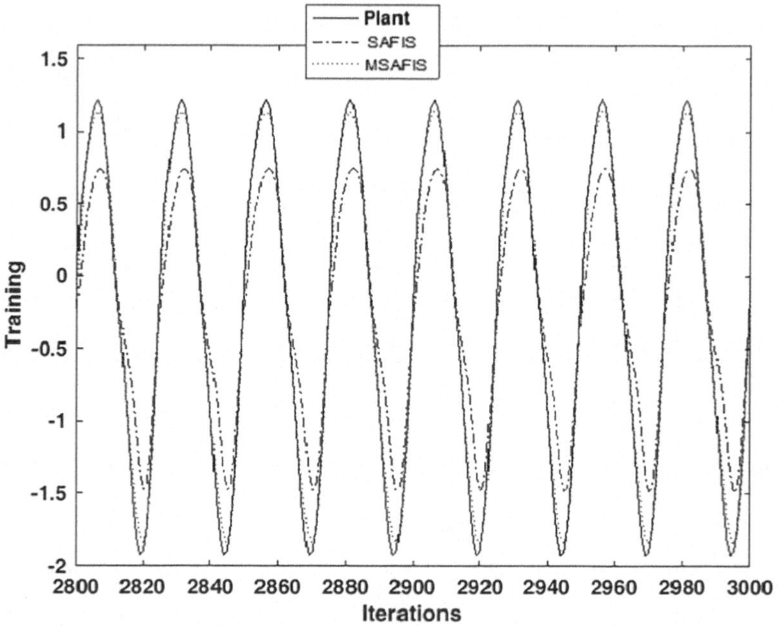

Fig. 11.5 Training for Example 1

Table 11.1 Comparisons for Example 1

Strategy	Training MSE
SAFIS	0.1740
MSAFIS	0.0730

of the first reaches better the plant signal than the signal of the second. Then, the MSAFIS is the best option for the plant modeling in Example 1.

5.2 *Example 2*

The plant of Example 2 is expressed in the next equation:

$$\gamma(k) = 0.3\gamma(k-1) + 0.6\gamma(k-2) + f(\upsilon(k-1)),$$
$$f(\upsilon(k-1)) = 0.6\sin(\pi\upsilon(k-1)) + 0.3\sin(3\pi\upsilon(k-1)) + 0.1\sin(5\pi\upsilon(k-1)),$$
$$\upsilon(k-1) = \sin\left(\frac{2\pi(k-1)}{200}\right), \tag{11.30}$$

The nonlinear plant of Eqs. (11.1) and (11.30) where inputs are $\chi_1(k) = \gamma(k-1)$, $\chi_2(k) = \gamma(k-2)$, $\chi_3(k) = \upsilon(k-1)$ and the output is $\gamma(k) = \gamma(k)$. The data of 3000 iterations are used for the training.

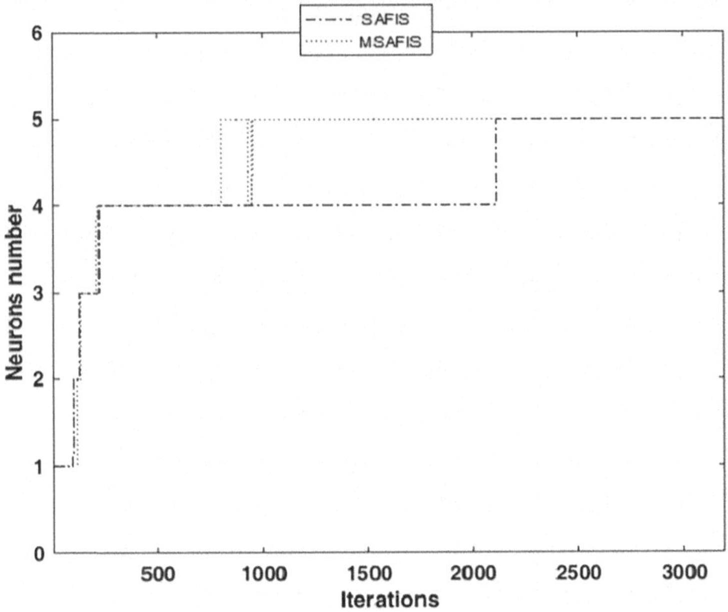

Fig. 11.6 Neurons number for Example 2

Terms of the SAFIS algorithm [11] are $N = 3$, $\eta_s = 0.99$, $K_s = 1$, $\varepsilon_{s,\max} = 5$, $\varepsilon_{s,\min} = 0.5$, $\gamma_{s,g} = 0.01$, $\gamma_{s,p} = 0.001$, $g_{s,e} = 0.01$, $b_{s,2} = 0.2$.

Terms of the MSAFIS algorithm [13] are $N = 3$, $\eta_c = 0.99$, $K_c = 1$, $\varepsilon_{c,\max} = 5$, $\varepsilon_{c,\min} = 0.5$, $\gamma_{c,g} = 0.01$, $\gamma_{c,p} = 0.001$, $g_{c,g} = 1$.

Figures 11.6, 11.7, 11.8, and 11.9 show the comparisons for the neurons number, the generated neurons, MSE convergence, and training of the SAFIS and MSAFIS. The training MSE comparisons of (11.28) are shown in Table 11.2.

From Figs. 11.6 and 11.7, it is observed that both algorithms reach the same neuron number. From Fig. 11.8 and Table 11.2, it is observed that the MSAFIS has better convergence than the SAFIS because the MSE is smaller for the first. From Fig. 11.9, it is observed that the MSAFIS improves the SAFIS because the signal of the first reaches better the plant signal than the signal of the second. Then, the MSAFIS is the best option for the plant modeling in Example 2.

Remark 11.8 Take into account that SAFIS and MSAFIS have the purpose to use the least required neurons to get a satisfactory modeling. SAFIS and MSAFIS of Example 2 in this chapter are compared with the well-recognized ANFIS algorithm of Example 3 in [27]. While ANFIS uses seven neurons to get a satisfactory result, SAFIS and MSAFIS use five neurons denoted in Fig. 11.6. This result shows the SAFIS and MSAFIS use a less number of neurons than the ANFIS. Thus, the SAFIS and MSAFIS are more compact than the ANFIS for the modeling.

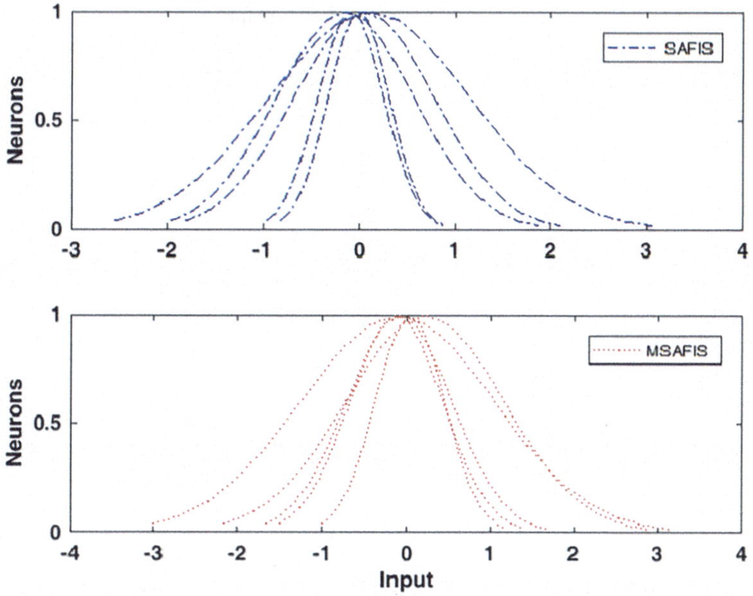

Fig. 11.7 Generated neurons for Example 2

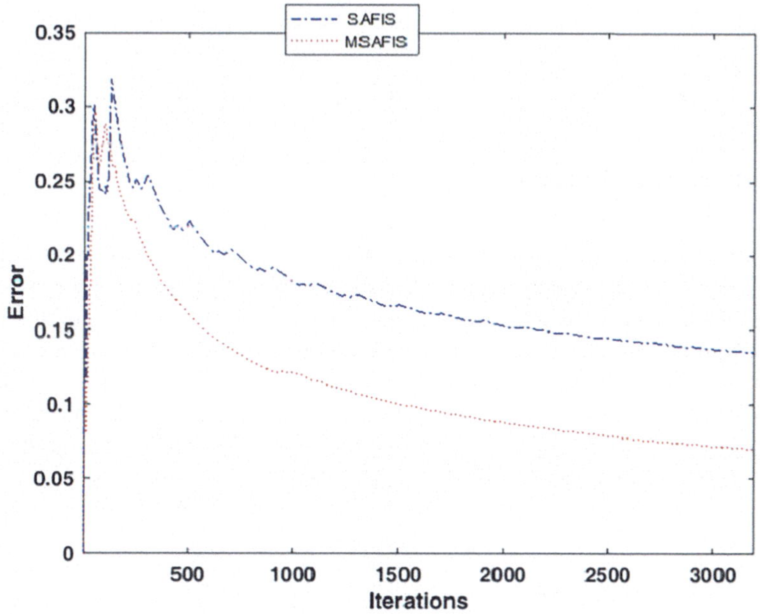

Fig. 11.8 MSE convergence for Example 2

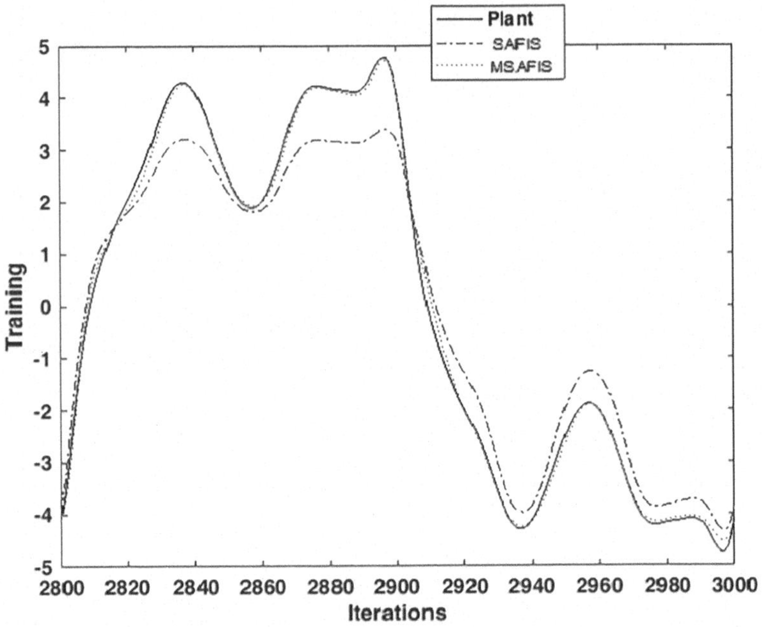

Fig. 11.9 Training for Example 2

Table 11.2 Comparisons for Example 2

Strategy	Training MSE
SAFIS	0.1349
MSAFIS	0.0705

6 Concluding Remarks

In this chapter, the error convergence of the SAFIS and MSAFIS was ensured. Utilizing two different examples, the MSAFIS produced higher accuracy compared to the SAFIS. It is worthwhile to mention, because as MSAFIS and SAFIS are based on online modeling, they can handle datasets of any size. They can also be applied in machine failure detection or management of the life cycle. Here they were successfully applied for the modeling of nonlinear plants. In the future, the error convergence of other evolving intelligent networks will be analyzed, or the SAFIS and MSAFIS will be applied in a prognostic health management plant.

References

1. E. Lughofer, On-line active learning: a new paradigm to improve practical useability of data stream modeling methods. Inf. Sci. **415–416**, 356–376 (2017)
2. E. Lughofer, M. Pratama, I. Skrjanc, Incremental rule splitting in generalized evolving fuzzy systems for autonomous drift compensation. IEEE Trans. Fuzzy Syst. **26**(4), 1854–1865 (2018)
3. S. Mohamad, A. Bouchachia, M. Sayed Mouchaweh, A bi-criteria active learning algorithm for dynamic data streams. IEEE Trans. Neural Networks Learn. Syst. **29**(1), 74–86 (2018)
4. M. Pratama, G. Zhang, M.J. Er, S. Anavatti, An incremental type-2 meta-cognitive extreme learning machine. IEEE Trans. Cybern. **47**(2), 339–353 (2017)
5. M. Pratama, E. Lughofer, M.J. Er, S. Anavatti, C.P. Lim, Data driven modelling based on recurrent interval-valued metacognitive scaffolding fuzzy neural network. Neurocomputing **262**, 4–27 (2017)
6. M. Pratama, P.P. Angelov, E. Lughofer, M.J. Er, Parsimonious random vector functional link network for data streams. Inf. Sci. **430–431**, 519–537 (2018)
7. F. Serdio, E. Lughofer, A.C. Zavoianua, K. Pichler, M. Pichler, T. Buchegger, H. Efendic, Improved fault detection employing hybrid memetic fuzzy modeling and adaptive filters. Appl. Soft Comput. **51**, 60–82 (2017)
8. H. Toubakh, M. Sayed Mouchaweh, Hybrid dynamic data-driven approach for drift-like fault detection in wind turbines. Evol. Syst. **6**, 115–129 (2015)
9. H. Toubakh, M. Sayed Mouchaweh, Hybrid dynamic classifier for drift-like fault diagnosis in a class of hybrid dynamic systems: application to wind turbine converters. Neurocomputing **171**, 1496–1516 (2016)
10. Y. Zhang, M.J. Er, R. Zhao, M. Pratama, Multiview convolutional neural networks for multidocument extractive summarization. IEEE Trans. Cybern. **47**(10), 3230–3242 (2017)
11. H.J. Rong, N. Sundararajan, G.B. Huang, P. Saratchandran, Sequential adaptive fuzzy inference system (SAFIS) for nonlinear system identification and prediction. Fuzzy Sets Syst. **157**(9), 1260–1275 (2006)
12. H.J. Rong, N. Sundararajan, G.B. Huang, G.S. Zhao, Extended sequential adaptive fuzzy inference system for classification problems. Evol. Syst. **2**, 71–82 (2011)
13. J.J. Rubio, A. Bouchachia, MSAFIS: an evolving fuzzy inference system. Soft Comput. **21**(9), 2357–2366 (2017)
14. C.K. Ahn, P. Shi, R.K. Agarwal, J. Xu, L∞ performance of single and interconnected neural networks with time-varying delay. Inf. Sci. **346–347**, 412–423 (2016)
15. G. Andonovskia, P. Angelov, S. Blazic, I. Skrjanc, A practical implementation of robust evolving cloud-based controller with normalized data space for heat-exchanger plant. Appl. Soft Comput. **48**, 29–38 (2016)
16. A. Bayas, I. Skrjanc, D. Saez, Design of fuzzy robust control strategies for a distributed solar collector field. Appl. Soft Comput. **71**, 1009–1019 (2018)
17. A.K. Das, S. Sundaram, N. Sundararajan, A self-regulated interval type-2 neuro-fuzzy inference system for handling nonstationarities in EEG signals for BCI. IEEE Trans. Fuzzy Syst. **24**(6), 1565–1577 (2016)
18. V. Subbaraju, S. Sundaram, S. Narasimhan, Identification of lateralized compensatory neural activities within the social brain due to autism spectrum disorder in adolescent males. Euro. J. Neurosci. **47**(6), 631–642 (2018)
19. Y. Pan, T. Sun, Y. Liu, H. Yu, Composite learning from adaptive backstepping neural network control. Neural Networks **95**, 134–142 (2017)
20. Y. Pan, M.J. Er, T. Sun, B. Xu, H. Yu, Adaptive fuzzy PD control with stable h∞ tracking guarantee. Neurocomputing **237**, 71–78 (2017)
21. T. Sun, Y. Panb, C. Yang, Composite adaptive locally weighted learning control for multi-constraint nonlinear systems. Appl. Soft Comput. **61**, 1098–1104 (2017)

22. G. Rajchakit, R. Saravanakumar, C.K. Ahn, H.R. Karimi, Improved exponential convergence result for generalized neural networks including interval time-varying delayed signals. Neural Networks **86**, 10–17 (2017)
23. R. Saravanakumar, M. Syed Ali, C.K. Ahn, H. Reza Karimi, P. Shi, Stability of Markovian jump generalized neural networks with interval time-varying delays. IEEE Trans. Neural Networks Learn. Syst. **28**(8), 1840–1850 (2017)
24. J.O. Escobedo, V. Nosov, J.A. Meda, Minimum number of controls for full controllability of linear time-invariant systems. IEEE Latin Am. Trans. **14**(11), 4448–4453 (2016)
25. A. Grande, T. Hernandez, A.V. Curtidor, L.A. Paramo, R. Tapia, I.O. Cazares, J.A. Meda, Analysis of fuzzy observability property for a class of TS fuzzy models. IEEE Latin Am. Trans. **15**(4), 595–602 (2017)
26. J.J. Rubio, Error convergence analysis of the SUFIN and CSUFIN. Appl. Soft Comput. **72**, 587–595 (2018)
27. J.S. Roger Jang, ANFIS: adaptive-network-based fuzzy inference system. IEEE Trans. Syst. Man Cybern. **23**(3), 665–885 (1993)

The manufacturer's authorised representative in the EU is Springer Nature Customer Service Centre GmbH, Europaplatz 3, 69115 Heidelberg, Germany. If you have any concerns regarding our products, please contact ProductSafety@springernature.com

Printed and bound by CPI Group (UK) Ltd, Croydon, CR0 4YY
01/10/2026
02225956-0006